UNDERSTANDING BRITAIN'S WILDLIFE

UNDERSTANDING BRITAIN'S WILDLIFE

Text and photographs by

Ian Beames

David & Charles

(Page 2) A mute swan in the fringe of a marsh dyke. The ducks, geese, grebes and swans normally associated with the open water of large lakes and ponds build their nests in the vegetation on its edges or in the ditches, dykes and tussocks of the open marsh. Mute swans are sometimes seen in estuaries and seaside harbours where there are rich pickings from holidaymakers.

The publishers gratefully acknowledge the generous contribution made by BP towards the production of this book. BP supports organisations, both in the UK and overseas, working on practical conservation projects, protecting threatened habitats, improving environmental education and awareness, and tackling the fundamental issues affecting the environment worldwide.

Examples of such support include: sponsoring a marketing programme to raise the profile and the membership of the British Butterfly Conservation Society; a joint venture with the RSPB to develop and demonstrate heathland management techniques; sponsoring the Council for Environmental Education's national youth training programme to encourage and implement environmental education throughout the youth service in the UK; and funding a two-year 'National Village Appraisals' project to establish the needs and views of local people on the future development of their rural communities.

The views expressed in this book are those of the author, and not necessarily those of BP.

All photographs by the author

British Library Cataloguing-in-Publication Data
Beams, Ian
  Understanding Britain's Wildlife.
  I. Title
  574.941

ISBN 0-7153-9861-X

Typeset by ABM Typographics Ltd, Hull
and printed in The Netherlands
by Royal Smeets Offset, Weert
for David & Charles plc
Brunel House   Newton Abbot   Devon

CONTENTS

INTRODUCTION

The best way to learn about Britain's wildlife is to get out and see it. With a little commonsense most of Britain's wildlife can be seen in complete safety; only in the mountains and on the cliffs is there any real danger, although I have fallen 20ft (6m) out of a tree in wildest and most dangerous Surrey! Always remember that, in the hills, thick mists and blizzards can quickly turn an easy walk into a real struggle for survival, so jeans and trainers are not appropriate: a pair of strong walking boots, a good waterproof jacket, and plenty of extra warm clothing and food are all prerequisites. Maps and compass are essential, and you must know how to use them; and take an ice-axe on any high hillside where there is snow.

Sensible clothing is a must for all wildlife work, in a quiet colour as a bright orange anorak is just as conspicuous to any fox, deer or bird as it is to us. The equipment you will need depends upon your particular interest. A pair of binoculars is essential, though do try to avoid very heavy ones as they are difficult to hold up for long. A good quality pair of either 8 × 30 or 10 × 40 glasses will cost upwards of £150 (1991), and will last for many years; a cheaper pair will neither last so long nor give such a bright image. Carry a × 10 lens in your pocket to examine the smaller objects – you will be amazed by the beauty, structure and symmetry of many tiny, fragile scraps of life.

But be warned! Wildlife photography is a bit like a drug – once hooked you may never be able to give it up, and camera equipment can run away with your money. A good beginner's kit would include one SLR body, two zoom lenses (28–80mm and 70–210mm), a 2× converter to convert the 210mm lens into a 420mm telephoto, and a close-up lens for insect and flower photography. Two medium flashguns are also needed, as well as a tripod and a cable release. At the lower end of the market all this will cost about £500 (1991).

This book tries to show the importance of conservation. It is ardently critical of the ravaging of Britain's countryside by the developer who probably doesn't live anywhere near the urban monstrosities he has created – and by the farmer, who is turning the countryside in which he does have to live into a regimented, chemical-filled environment. Intensive farming methods run the risk of leaving ecologically barren countryside in their wake. One by one the hedgerows are uprooted and with them a whole swathe of plants and animals disappear.

Habitat destruction is therefore the biggest threat to our wildlife – far more so than the more direct means which man employs to control or destroy the natural world around him. It depends on each one of us to stop the destruction of our environment, from back-street wildlife parks threatened with suburban in-filling, to the great expanse of the Sutherland Flows.

Ian Beames, Epsom, Surrey, 1991

Country foxes, like this alert young animal, have much larger home ranges than city foxes because food is more difficult to find. The best time to see foxes in the countryside is early in the morning or at dusk, especially in places where rabbits are numerous. If the fox becomes suspicious, stand stock-still – as soon as you move and show your human shape, the fox will run.

CHAPTER 1
WOODLANDS

TWELVE THOUSAND years ago the ice-caps which covered Britain retreated northwards, leaving a land of cold mountains, icy moorland and arctic tundra. For thousands of years the ice had covered the land as far south as the Wash on the east, and South Wales in the west. South of its limits, mammoths and reindeer roamed the tundra, while on the few exposed high peaks there must have been some bird species which could stand the arctic conditions, like ptarmigan, or dunlin, an arctic wader. At that time Britain was still attached to the mainland of Europe because the ice – several miles (kilometres) thick in places – had locked up huge quantities of water.

As the climate improved, and the ice began its slow retreat, plants and animals pushed further northwards; as the tundra gave way to more amenable conditions, so forest trees were able to grow, spreading from the rest of Europe. The birch was probably the first to take hold, followed by the Scots pine which soon became dominant, with huge forests spreading over much of the country. Scattered remnants of these forests still survive today in the Scottish Highlands where the climate is harsh and sub-arctic.

The climate became increasingly warmer and the northward spread of the forests continued. By 5000BC much of lowland Britain was clad in a mixed deciduous woodland, dominated by oak, also with hazel, lime, ash and elm, and pine had established itself further north. There were two species of oak: pedunculate oak on the heavy, water-retentive soils of the lowlands, and sessile oak on the more acid soils of the north and west. However, the harsher Scottish climate prevented oak from establishing itself in the Highlands.

Knowledge of this woodland succession is revealed from studies of pollen remains in peat deposits. Pollen grains are extremely resilient, and because each species has a distinctive shape and pattern, it is easy to identify them under a microscope. Thus it is possible to estimate the relative abundance of tree species at any particular time in the past. These studies show that as the oak became established, so did other important trees such as hazel, with alder in the marshland, and ash on the drier northern limestones. Lime was also widespread but because it is insect pollinated it is under-represented in any pollen analysis.

With the trees came the animals which lived in them: black grouse with the Scots pine, badgers and wild boar with the oak. Excavations carried out at Star Carr in Yorkshire have revealed that 9,600 years ago elk, beaver, red and roe deer, fox, badger and wild boar had all arrived in Britain. After the Ice Age all the other typically woodland species –

PREVIOUS PAGE
A beech wood in summer. Planted some two centuries ago for timber, it is now used as amenity woodland. When trees started to be felled for building and other purposes, the new growth from some deciduous species produced poles which were ideal for fencing, wattles, charcoal and fuel. This coppicing in turn allowed the growth of vegetation from dormant seeds (due to the increased light availability) which encouraged birds and mammals to seek its shelter.

LEFT
The woodlands of the New Forest, managed by man for centuries, link the modern forest with the ancient wildwood. Examination of pollen grains in peat deposits reveals the relative abundance of tree species which have existed in the past; in 5000BC much of lowland Britain was covered in mixed deciduous woodland. Ditches are often signs of past management of wood pasture, as they were used to keep stock within a given area. The animals would eat the ground flora and gradually turn the area into grassland.

including man – also made their way into these lands.

As the ice melted, so the sea-level rose and flooded the low-lying areas, creating wide channels between Ireland and Scotland – even now only some 200ft (65m) deep – and separating what became the British Isles from the European mainland; at this point the English Channel was still on average only some 60ft (20m) deep. However, the flooding of the Irish Sea prevented many animals from reaching Ireland, which is why the mole, the common shrew, the dormouse, snakes, and all voles except the bank vole are not found here today.

Slowly, early man changed from a hunter-gatherer to a primitive farmer; about 5,000 years ago, he started to fell the forests, first with large stone axes and then – far more efficiently – with bronze. The Celtic tribes brought iron to Britain, and the land rumbled to the crash of falling trees as forests were cleared to build and make way for prehistoric farmsteads. Romans, Danes and Anglo-Saxons, all were farmers, and by the time William of Normandy strode ashore in 1066 England was already a pastoral land, a patchwork of fields and woods; and so it has remained almost to the present day.

The pine forests of Scotland managed to survive well into the Iron Age, even when the iron industry was expanding so rapidly, but already by Norman times England was no longer a place of great forests. The Forest Law of Norman kings made provision for nearly a third of the land to be used for the court's hunting pleasure, and this probably preserved the few remaining ancient forests.

There is evidence that as early as Roman times man began to realise that some management of the forest was needed if it was to survive at all. Once a village had built its houses, its prime need was for light wood, primarily for fuel, and not for main timber. Early man soon discovered that most broad-leaved trees when cut to the ground do not die but sprout a number of poles. He therefore applied the principle to the woods in his vicinity, cutting only some trees down, then allowing them to sprout poles. Areas of these forests were cropped every ten to twenty years to provide light timber for fencing, wattles, charcoal and fuel. Some trees were allowed to grow to full height;

these were known as 'standards' and provided a longer-term timber crop for building. This method has been used for many centuries, and is known as 'coppicing-with-standards'; its advantage is that the rest of the original wildwood can be left largely intact. Conservation of woodland is, therefore, no new phenomenon, although only very recently has it been considered necessary to safeguard it purely for its amenity and wildlife value.

Coppicing inevitably provided a wide variety of trees at different stages of growth, usually to the great benefit of the ground flora and fauna. In the first year or two after coppicing flowers would surface in great numbers from seeds that were previously deprived of light, but were now exposed to the sun again. Then, as the scrub grew taller, birds and mammals would seek its shelter to breed in abundance. Bradfield Woods in Suffolk, now a nature reserve, has been coppiced since the thirteenth century and contains at least 350 species of flowers and 42 different types of trees and shrubs.

'Wood pasture' consists of grassland and mature trees and was the result of a different system of ancient management. It is seen today in places like the New Forest in Hampshire or Richmond Park in Surrey. A coppice usually had a boundary in the form of a raised border and a ditch on the outside to keep the grazing animals *out*, whereas wood pasture had the ditch on the *inside* to keep the stock *in*. The idea was for the sheep, goats, cows and pigs to eat all the ground flora and turn the area into grassland, under the broken canopy of its larger trees.

Only after World War I, when nearly 600,000 acres (200,000ha) of ancient forest were felled, did the urgent need for timber in Britain become apparent. The Forestry Commission was established in 1919; after a further 400,000 acres (160,000ha) of old woodlands were felled in World War II, it began its revolutionary policy of establishing alien conifers, which grow faster than Britain's mighty oaks. It planted them on a massive scale.

Now, only 9 per cent of the land in Britain is forested, and all but 1 per cent of it with new conifer plantations. In Europe, only Holland, much of which consists of land reclaimed from the sea, has as little woodland as Britain. Germany still possesses a 25 per cent forest cover despite pollution by acid rain, and Japan, perhaps the most industrialised nation on earth, has a 60 per cent forest cover – the Japanese apparently revere their trees and prefer to get their timber from other countries! The sad fact is that once an ancient wood is destroyed it has almost certainly gone for ever: no one so far has ever been able to re-create the particular assemblage of trees and plants, gathered by nature down the centuries, which comprises a wildwood. It is a tragedy that between 1945 and 1985 we cut down over one-third of our ancient woodlands, primarily to replant with conifers. The ancient wildwood is therefore in need of urgent protection.

Deciduous Trees in Britain

Most native British trees are deciduous, although today our forest cover consists mainly of fast-growing conifers. Nearly all Britain's woodlands have been managed and modified by man, and very few can be traced back, in unbroken succession, to the ancient wildwood which was established after the Ice Age.

Oak is one of the best known British trees, and, as we have seen, there are two species: the pedunculate oak which is primarily found in heavy lowland soils, and the sessile oak, more often found on the lighter sandy soils of the north and west and in areas of higher rainfall. The two may grow together, as in the Wyre Forest near Birmingham, but can be distinguished by their leaves and acorns: the pedunculate oak has no stalks to its leaves, and long stalks for its acorns; whereas the sessile oak (the name means 'stalk-less') has no stalk to the acorns, and a short stalk for its leaves. Oaks provide the strongest timber and can survive many fungal and insect attacks virtually unscathed. The oak leaf-roller moth is their most notable assailant and its larvae will sometimes completely strip a tree of its leaves. However, the oak can survive such an attack with only a small increase in its annual growth. Oaks have been known to reach 100–130ft (30–40m) in height, and in a good year a large tree is capable of producing a bumper crop of

about 70,000 acorns. They often survive to a great age, living for longer than any other native British tree, and those which have been protected in long-established parklands such as Windsor or Richmond, or in the New Forest, may be eight or nine hundred years old.

The ash is the tree most frequently associated with oak on lowland chalky or clay soils. It is tall and graceful, and easily recognised – in spring its many separate leaflets allow far more light into the wood than do the leaves of the beech (which also favours a chalk and limestone soil); and in winter the hard black buds on its smooth grey twigs will quickly identify it. Its timber is very tough, but easy to bend, so it is much in demand for furniture-making. It grows best on a loam soil (clay-sand mixture), and, together with wych elm, is the predominant tree in woods from the White Peak (a district of the Peak District National Park) northwards. Ash trees may live for up to two hundred years and can reach a height of 130ft (40m).

In the south, the clay soils on the northern slopes of the downs support oak woods whereas chalkland areas are usually clad in beech. The tall, smooth, grey, fluted trunks of beech are easy to identify at any time of the year, and in early May the trees' newly opened leaves really make the beech wood a joy to walk in, the air seemingly filled with lemon-green light. However, once fully opened the leaves cast a deep shade so that not many plants are able to grow and, therefore, a beech-wood floor is usually bare. A healthy beech will live for some two hundred years and may be 100ft (30m) tall to the crown. A heavy crop of nuts is produced about every four years; these are known as 'mast years' and vastly increase the chances of survival for many animals and birds through the winter. Like the beech, a number of smaller trees and shrubs also prefer the sweet soil

Horse chestnut leaves in autumn. As the sap in the leaves dries up and the green chlorophyll is dispersed, colourful tannins fill the dying leaves. The nuts from chestnuts and hazels are foraged by squirrels, mice, badgers and deer.

of the chalk downs: field maple, privet, guelder rose, spindle and dogwood all grow best on these lime-rich soils.

In some places the dominant tree on chalk is not beech: it may be yew, a native conifer (as at Kingley Vale in Sussex where the best yew woods in Europe can be found) or box, primarily at Box Hill, near Dorking in Surrey. Box is one of our few evergreen broad-leaved trees and is particularly valuable for its superb hard timber, used for drawing instruments and wood engravings.

Many of these trees provide abundant food for numerous bird species – acorns, beech nuts and yew berries in autumn, and huge quantities of insect larvae in summer.

There is yet another group of nut-bearing trees, which rarely proliferates so as to produce a full-sized woodland but nevertheless still plays an important part in the wildlife cycle: the horse chestnut and the sweet chestnut, both foreign species, the first introduced some 400 years ago from the Balkans, the latter during the Iron Age and supplemented by the Romans. Both species have become common in certain places in the south, and both provide great quantities of nuts in autumn. So does our much smaller native tree, the hazel, whose nuts are foraged by squirrels, mice and even badgers and deer.

Hazel is now found mostly as a shrub in woodlands of various types, but it was once a favourite tree for coppicing, producing scores of poles for fencing, hurdles and even for basketwork and walking sticks. Hazel coppice 'stools' (stumps), if cut regularly, may survive for eight or nine hundred years. Hazel produces pollen from its catkins in early spring, which attracts many insects; if allowed to grow to tree-size, the bush may reach 25–30ft (7.5–9m) and then of course it produces far more pollen. However, hazel woodland has declined dramatically, as uses to which it was put in the old days have been superseded by plastics and other modern materials. From 500,000 acres

(200,000ha) in 1950 there are now fewer than 12,000 acres (5,000ha) left.

In marshy areas and along riverbanks the characteristic trees are the alder and the willow. The alder is a tree of our ancient wildwood, and is found throughout the country. It may grow to 60ft (20m) or more as an isolated tree, and is easily recognisable by its long red catkins in spring and the bunches of tiny cones often present throughout the year. However, normally it forms low-level, dense alder swamps called alder 'carr', a word derived from the old Norse word *kjarr* meaning a 'thicket'. Alder carr is a stage in the succession from wetland to dry, usually running into birch woodland as the soil level rises above the water table and dries out. Alder lives only where there is adequate water; its roots contain nodules which help to 'fix' nitrogen and its seeds float easily to colonise further areas.

The willow is the other tree most characteristic of wetlands. Several species are very familiar, especially the weeping willow, and the white willow, which is frequently used for pollarding. Pollarding is similar to coppicing but is carried out at about 8–10ft (2.5–3m) above the ground instead of at ground level, thus keeping the growing crown out of the reach of livestock and providing stout branches which are used for fuel.

The deciduous trees of the uplands include the birches – 'silver' and 'downy' – the aspen and the rowan. Birch is a coloniser, and is found in many woodlands throughout Britain. It produces thousands of tiny seeds which are easily transported by the wind, and this enables it to colonise any untouched ground, although it seems to grow best on acid soils. The silver birch is usually found in drier situations than the downy birch, which is the more common tree in Scotland. Birch seeds provide food for a variety of birds, particularly finches. Birches are relatively short-lived, and rarely survive for more than eighty years.

Rowan bears a superficial resemblance to ash (hence its other name of 'mountain ash'). It is widespread in the Highlands, and can be found as far up as the 3,000ft (900m) contour. Its autumn berries are eaten by many birds. The aspen is also a northern species and is one of the most striking contributors to the golden glory of the Highland autumn.

The Ecology of Deciduous Woods

Britain's broad-leaved woodlands vary in age from ancient wildwood many centuries old, to oaks planted comparatively recently. 'Primary' woodland can be traced right back to the original post-glacial forests, while 'secondary' woodland has been planted by man in the last few centuries. The variety of wildlife in a wood is directly related to its age and determining how old a wood may be can be difficult, especially more recent secondary woodland. But where there is a wide variety of common spring flowers like bluebell, wood anemone, yellow archangel, bugle and wood sorrel the wood is likely to be many centuries old. Shrubs from an ancient

wood are unlikely to colonise secondary or new woodland nearby because many – for example, hazel and field maple, both with large seeds – are very poor colonisers; they do not even move into adjacent hedges despite maybe 400 years of opportunity.

Broad-leaved woodland supports a wider range of plants and animals than any other habitat outside the tropics, although the survival of many of its inhabitants is now dependent upon how sensitively it is managed. Any habitat depends upon the continuing life-cycle of its plants and animals, a system which is complex and interdependent. So many people who walk through a wood have little idea of the intense activity going on around them; the woodland scene appears to be one of such peace and tranquillity.

Britain's highly seasonal climate dictates the life-cycle of all its plants and animals. The warm, moist conditions of spring and summer bring woodland trees and shrubs into leaf, and the energy from the sun is stored in them by photosynthesis. Leaves in turn will nourish the countless hordes of herbivores from tiny caterpillars to rabbits, and these creatures sustain a range of predators from wolf spider to red fox.

As autumn moves into winter and the weather deteriorates, many species of plants and animals hibernate or become quiescent so as to survive the hard conditions. Tree roots can only absorb water with difficulty from frozen ground, and so they shed their leaves to save further moisture loss. It is interesting to note that as trees in the tropical rain forests do not have this problem of climatic variation they are largely evergreen, and shed only their dead leaves throughout the year.

On the forest floor, the layers of fallen leaves provide insulation for a host of woodland species.

Sulphur tuft fungi are common in deciduous woods in autumn and are often found growing on dead tree stumps. There is a wide variety of species of fungi in Britain, providing a rich food source for insects and small mammals and thus playing an important part in the woodland eco-system.

The mass of fungal threads and multitude of tiny creatures such as springtails and woodlice all play a vital part in the natural cycle as they help to break down the leaves and any other dead organic matter. This enables chemicals released in the process of decay to be reabsorbed by the soil and so recycled in the woodland eco-system.

Woodland has an extra dimension that other habitats do not have: that of height, and this allows different species to live at different levels without having to compete. The four obvious layers are the canopy, the shrub layer, the field or herb layer of flowers and the ground layer of leaf litter. There may also be two sub-layers – one of young trees, the other of taller grasses and flowers – lying below the level of nettles and branches but higher than the herb layer. Some woodland trees produce such a thick canopy in spring and summer that the excessively shady conditions prevent the development of the shrub and field layer altogether – beech and yew woods are prime examples. Oak and ash woods on the other hand let in more light and all four layers are able to thrive.

Open areas are vital to the survival of many woodland species. In the ancient wildwood such areas would be created each time a tree fell; this let the light in, encouraging the herb layer and producing natural glades. Grazing would encourage these, too. The edges of these open areas, and the natural borders of a wood are important: unlike the straight manicured periphery of an artificial wood, a natural woodland border merges gradually into the surrounding countryside, its greater variety of vegetation encouraging many different kinds of plants and animals. Here, species typical of all three habitats will be found: of wood, open country, and of the 'edge' itself.

The lowest layer on the forest floor teems with life: spiders, mites, worms, springtails, woodlice, beetles, slugs, ants and many more all help to break down and recycle the fallen material of the forest. Here, mice and voles live in runs just below the surface layer and feed on both fallen fruits and insects. Fungi and bacteria break down the complex litter yet more so that it can be absorbed by the soil, its nutrients then ready to succour the next generation of trees and plants. Provided the climate is relatively stable this woodland cycle will continue indefinitely.

It is fascinating to watch this cycle unfold. A good time to start is after leaf fall in late autumn, as it is easier to identify the freshly fallen leaves and the bark patterns of the trees above. There is more light in the wood after the dark, shady conditions of high summer, and a sharp autumn morning with its low yellow light has a particular magic. The birds are sleek and plump having put on a layer of fat to sustain them through the winter cold, and throughout the wood, stripped down for winter, wildlife is more easily seen. The structure of the wood is also laid bare, whether it is regenerating naturally with seedlings and saplings of all ages, or coppiced, with many slender trunks growing from an ancient central stump.

With the onset of winter proper the wood becomes increasingly quiescent, sleeping out the cold, its life-patterns just ticking over. Mammals only forage for food when necessary; insects hibernate in cracks and crevices or deep in the leaf litter. Only the birds are really active, hunting for every available scrap of food to fend off the cold of night.

A few short weeks after the turn of the year the winter woods will assume an almost imperceptible touch of green; blackbirds chase each other through the still bare branches, while a cock chaffinch sings its first full song. Spring is about to burst forth. The cycle of the seasons has come full circle once again.

Coniferous Trees in Britain

Conifers, or softwood trees, form a distinct and important group in the world of forestry, because they produce top-quality industrial timber, yet will grow very quickly on poor soil. They can also withstand a harsh climate. Hardwoods, or broad-leaved trees, usually have a much slower growth rate; and it is a long time before their timber is marketable.

There are only three conifers native to Britain – the Scots pine, the juniper and the yew, although scores of other species have been planted here. About a dozen of these (mostly species from the

Yew berries, highly poisonous to man, provide a feast for thrushes in the chill of winter. The yew, one of only three conifers native to Britain along with the Scots pine and the juniper, grows well in the wild on chalk downland and limestone soils. Yews rival oaks as the oldest trees in Britain; they are often found in churchyards, and some are believed to be over a thousand years old.

New World) have been used extensively in forestry plantations, to provide quick-growing trees for timber production. Conifers nearly always have narrow needle- or scale-like leaves, scaly buds, a pattern of branches arranged with almost geometric precision, and a resinous fragrance; they are usually evergreen, with the notable exception of the larches. The male and female flowers are always borne separately, but often on the same tree; they are always wind pollinated and the fruit is usually formed in a woody cone, although yew and juniper have fleshy berries.

In order to identify a particular species of tree it is necessary to establish to which genus (grouping of similar species) it belongs – for example, fir, pine, spruce, hemlock, cedar and larch. For positive identification, examine the needle arrangements on the adult or mature branches, which are always distinctive. Larches have needles only in spring and summer, and these are set in clusters of twenty or more from a single woody knob; junipers have sharp-pointed needles set in groups of three all round the stem; pine needles are usually long and in groups of two, three or five, but with a

sheath at the base; hemlock needles are soft and yew-like and of varying lengths, but grow independently. Cedars have whorls of leaves on small spurs; spruce, Douglas and silver fir can all look similar, but spruce needles have a tiny 'peg' which is left on the tree when they fall naturally, but which comes off with the needle if this is pulled away; Douglas fir needles have a minute stalk, and the terminal bud on the shoot is slender, brown and papery in texture; whereas the silver fir has a blunt, round bud.

The native Scots pine is easily identified, first by its paired 'pine' needles which are short (about 1½in (4cm) long) and blue-grey in colour, and also by its distinctive orange and grey bark pattern. It is only found in its wild state in the north, but it has been planted everywhere and will easily grow from seed on sandy heathlands all over the country. The golden male flowers shed clouds of pollen in May, fertilising the tiny female flowers which take two years to grow into mature green cones. The Scots pine can live for up to 400 years and may reach 130–60ft (40–50m) in height. Old mature trees can usually be identified easily by their rugged outline. The ancient native 'Caledonian forests' of Scotland are composed of this tree, and the best examples can be found in Strathfarar, Glen Affric, Speyside and Rannock Forest; there are also impressive plantations at Thetford Chase in Norfolk, Cannock Chase in Staffordshire, and in the New Forest.

Three other pines which have been extensively planted are the Corsican pine (a variety of the black pine of southern and eastern Europe), the Monterey pine and the lodgepole pine. The Corsican pine was introduced into Britain in 1792; today it is found in Britain in areas of low rainfall and high sunshine, in conditions which resemble its native climate as much as possible. Thus extensive plantations exist in East Anglia, along the east coast of Scotland, and on the coastal fringes of the north west. It was often planted around the turn of the century in coastal shelter-belts to stabilise sand dunes. Its greyish-green needles are usually 3in (8cm) long and distinctively twisted; the bark is grey, never orange as in the Scots pine.

The Monterey pine, introduced from the Californian coast in 1833, prefers the wetter, milder climates of Britain's west coasts and is now the common shelter-tree of south-west England. There are extensive plantations in southern Britain, where some of the very large mature trees may reach 100ft (30m) in height – much larger than in their native California. The needles grow in threes, and are long and bright grass-green; the cones may stay on the tree for fifty years.

The lodgepole pine has proved most successful in the west of Britain, and will grow on poor soil as long as it has a high rainfall. It comes from the damp west coast of North America and its name originates from the native Indian population who used its straight stems to support their wigwams. Its needles are 1½–2½in (4–6cm) long and of a mid-green tint; the bark divides into small square brown plates, quite unlike the Scots pine.

The two common species of spruce in Britain are the Norway spruce (well known as the nation's Christmas tree) and the Sitka spruce, which is now the most widely planted of all the 'foreign' conifers used for forestry. The Norway spruce grows wild over most of northern and central Europe but was only introduced to Britain in the sixteenth century. Its needles are mid-green in colour and not as

sharp as those of the Sitka; its bark has a rusty tint – hence its German name of 'red spruce'. The long, distinctive, cylindrical cones are about 6in (15cm) long and ripen in autumn, hanging down from the tips of the branches. The tree grows quickly, producing timber known as 'white deal' or 'white wood' in the trade and a mature tree may reach 130ft (40m).

Sitka spruce may be an amazing success in forestry terms, but it is a disaster in conservation and wildlife terms as it has completely colonised many of the wild moorland bogs in the Highlands of Scotland. It comes from west-coast America and is like a Norway spruce in shape, but its needles have a blue or slate-grey tint and they are very sharp, which prevents it being sold as a domestic Christmas tree. It needs a high rainfall and grows quickly in huge plantations in the north and west, producing bulk white softwood which can be used for all sorts of purposes, but especially for paper.

The two larches, the deciduous conifers, are both introduced species: the European larch with smaller, bright green needles from the Alps in about 1620, and the Japanese larch with longer, blue-grey needles and red twigs from Japan. The timber crop from these trees is fairly light and they need more space than others such as spruce. European larch may grow to 130–50ft (40–45m), while the Japanese larch tends to be smaller, reaching an average height of about 120ft (37m).

Yew and juniper, both native British conifers, are usually found as 'artificially' planted, ornamental trees, although yew does grow well in the wild on chalk downland and on limestone soils; as does juniper, only a shrub, but one which may reach 16ft (5m) on good soils. Juniper is part of the shrub layer of the ancient Scottish Caledonian forest. Both the yew and the juniper bear berries: those of the yew are bright red, and the juniper's are blue. Yew trees rival oaks as the oldest trees in Britain and some isolated specimens in churchyards may be a thousand years old.

The two silver firs most often encountered are

Larches in autumn; the yellow needles of Britain's only deciduous conifer will soon fall to the woodland floor.

the grand fir and the noble fir. Both came originally from North America, and are planted less extensively than the other conifers. They can become very tall, nearly 200ft (60m) high; the grand fir has flattened needles which lie parallel to the bark, while those of the noble fir are upswept, especially on the side twigs. Douglas fir, much favoured by foresters before the mass plantings of Sitka spruce, also comes from west-coast North America, and may also grow to 200ft (60m).

Western hemlock, with its graceful, drooping branches, has bright green flexible needles of varying lengths which seem to crowd along the twigs in a somewhat random fashion. It grows well in partial shade and produces a fine timber yield, so is often planted beneath a growing crop of other trees.

Plantations, with trees all of the same type and age, are planted entirely for commercial purposes and only with ease of thinning and felling in mind, and not for the benefit of wildlife. Nor should forestry operations be seen as replacing the ancient wildwood which is such an invaluable habitat for plant and animal life.

The Ecology of Coniferous Woods

The life of an evergreen conifer forest is quite different from that of a deciduous wood because the leaves, or needles, stay on the trees throughout the year; dead needles may be shed at any season throughout the tree's life. A deciduous wood, and the life-cycle of its plants and animals, are regulated to a much greater degree by the changing seasons. Coniferous woods are usually places of cool shade, since hardly any light can reach the forest floor. Very few plants or shrubs find these conditions suitable for growth, and often the only visible life-form to be found are fungi.

With the notable exception of the native pinewoods of the ancient Scottish forests (which are more open and therefore full of shrubs and flowers, and accompanying birds and animals) most of our forests are conifer plantations inha-

bited by only a few specialised species of birds and insects. These plantations do, however, offer shelter to other creatures from outside their bounds, and in Scotland they have encouraged the spread of the red and roe deer, and birds such as sparrowhawks and long-eared owls.

Since most of our native trees are deciduous, one might assume that evergreen trees in Britain must face certain problems. Maintaining a leaf-cover through the winter means that the leaves must be well protected, not only against frost but also, perhaps surprisingly, against water loss. In cold areas, like the Highlands, water may be frozen and unavailable for long periods; in this respect the Arctic can be as much a desert as the Sahara. Conifer needles are well suited to minimise water loss because of their needle-shape and thick, hard, leathery surface. On the other hand, because the needles remain on the trees all year, they are more susceptible to pollutants than deciduous trees which are usually bare at times of highest precipitation in winter. This is also the time of year when sulphur dioxide increases in the atmosphere due to higher fuel consumption, resulting in the modern problem of 'acid rain'.

In some situations, where the soil is poor and the climate harsh, evergreens have the advantage over deciduous trees. It is in these places that they thrive. They can also grow all year round through photosynthesis, their leaves are tough and withstand insect attack better, and their resinous wood is able to survive fungal attack very well indeed.

Native Scots pine forests are found only in the central Highlands and provide a rich wildlife habitat. The trees grow well apart from each other, the sunny glades allowing unrestricted growth of dense ground cover including plants such as heather, juniper and various other berry-bearing shrubs. The older trees invariably have holes in them, and fallen and dead trunks remain *in situ*, so there is a variety of niches to suit all sorts of different animals and plants. Redstarts and tits breed in the tree holes, and the rare crested tit is also found exclusively in this habitat. Many warblers nest in the scrub layer, especially willow and wood warblers; eagles, hawks and falcons hunt the canopy where crossbill, siskin, tree pipit and chaffinch all thrive. Ground-nesting game birds include black grouse and the capercaillie. And mam-

The field vole forms a large part of the diet of many predatory birds and mammals such as the buzzard and the fox, and so plays an important part in the main food chain. Plant-life in the northern forests supports a wide variety of insects and herbivores; birds feed on the insects and fall prey to falcons and hawks, while marten, wildcat and fox feed on squirrels, voles and ground-nesting birds.

mals are numerous with red squirrel, roe and red deer and the rare pine marten and wildcat all taking advantage of the forest's seclusion and plentiful food supply.

The dense shrub and field layers in the native pine woods are nearly all calcifuge (lime-hating) plants, used to growing on acid, sandy soils. The rainfall is often high and mosses are luxuriant, with bilberry, crowberry and heather also growing in abundance. Only tough, competitive plants gain a foothold, such as tormentil, heath bedstraw and wavy hair-grass. Here and there in the bare areas, sometimes beneath tall heather shrubs, plants can be found growing which are specific to these woods: the rare twinflower, several species of wintergreen, and three small orchids – the coral-root, creeping lady's tresses and lesser twayblade.

In these ancient northern forests all the species which constitute the main food chains are represented and the predator–prey relationships are balanced and stable. Plant-life is varied and supports a variety of insects as well as herbivores such as field voles, wood mice and deer. In turn, birds feed on the insects, while at the head of the chain come the large predators: falcons and hawks feed-

ing on the birds and the voles; marten, wildcat and fox feeding on squirrels and voles, and ground-nesting birds such as ducks and grouse.

In a man-made conifer plantation, the ground cover is often dense while the trees are small, with real thickets of bramble and grasses. Usually, rabbit fences have been erected to protect the young trees. In the early stages of plantation development birds such as pipits, larks, warblers and other ground-nesting species take full advantage of the thick cover. However, they are gradually obliged to leave by the time the forest reaches about 15ft (4–5m) in height, as the branches shade out the shrubs and thickets beneath; in any case, at this time the lower branches of the plantation are normally 'brashed' (cut away) and the ground cleared to reduce fire risk. The voles that fed the short-eared owl nesting in the young plantation vanish too, also unable to find a living in the changing habitat, and eventually, after some ten to fifteen years, all that is left is a 'pole plantation' with little or no ground cover, in deep shade throughout the year, with few insects, no small mammals, no flowers, and very little invertebrate life. The slow decay of pine needles adds to the acidity of the soil, inhibiting many soil organisms such as worms and snails; and because there are no fallen trees and no tree holes there are none of the invertebrates which normally live in dead wood (these may comprise a third of the species present in a natural wood). Of the birds, only canopy residents such as goldcrests and siskins survive in any reasonable numbers.

In winter, however, deer and some bird species do use conifer plantations for shelter when conditions are hard. Birds in particular may find more insects among the pine needles than they would in the exposed, leafless branches of the nearby deciduous wood. But these inhabitations are short lived, and as the weather improves, the man-made forest reverts to its virtually deserted state.

Woodland Flowers

In spring, most deciduous woods are full of flowers. Snowdrops in February are quickly followed by primroses, violets, celandines and wood anemones in March. All these early woodland flowers are adapted to produce their blooms at this time before the trees develop their leaves. Light floods into the wood in spring, whereas in summer the dense leaf canopy blots out the sun, resulting in dim green shade down below. Pollination of these early flowers is assured because there are always a few warm days which wake up the hibernating insects, and these are drawn at once to the first flowers. Which woodland flowers become established and the extent to which they proliferate depends upon the nature of the underlying soil, the amount of light or shade which is available, and the overall climate of the area.

Pedunculate oak woodland on neutral or calcareous soils has a rich and colourful ground flora including celandine, primrose, wood anemone, wood and sweet violets, lady's smock, early purple orchid, dog's mercury, bugle, wood spurge, yellow archangel and bluebell. Hazel is the shrub most typical of the oak wood; ivy and honeysuckle are the woodland climbers. In summer, foxgloves and rosebay willow-herb fill the glades and clearings with colour, and enchanter's nightshade grows in abundance in the shade. The tall marsh thistles which thrive along the rides attract many varieties of summer butterfly.

The bluebell is particularly characteristic of British woodland and a bluebell wood in full bloom is quite the most brilliant of spectacles. Unfortunately, the bluebell does not easily colonise new woodland and so it is very seldom found in the plantations established by man in the last two or three centuries. Large areas of bluebells in a wood means that it has been woodland for many centuries.

In the western sessile oak woods the flora is dominated by species preferring acidic soils – as well as the bluebell, wood sorrel, foxglove, tormentil, cow-wheat, bilberry and heather are seen most frequently.

On the chalk soils in southern Britain the climax forest is likely to be beech, its most significant characteristic being the thick canopy which casts a deep summer shade once the leaves emerge early in May. This effectively suppresses the ground flora, except for those species particularly adapted to shady conditions such as dog's mercury, sanicle, sweet woodruff and dog violet. A beech wood is often rich in orchids – including fly and butterfly – and in various other helleborine species. Deadly nightshade is usually found in the clearings, as is nettle-leaved bellflower, and green and stinking hellebore – all appear to prefer the rather specialist conditions of the beech wood. Here too grow saprophytic plants such as bird's-nest orchid, yellow bird's-nest and, rarest of all, the ghost orchid, all living on the decaying leaf litter. Because they contain no chlorophyll, they do not depend on photosynthesis for survival and therefore thrive despite the lack of sunlight.

In the north and west where harder limestone replaces the softer chalk, ash is the more usual

The early purple orchid is the first of Britain's orchids to flower and is found from April to June in woodland, on roadside verges and grassland, particularly on chalk. It is occasionally paler in colour, sometimes almost white.

Primroses in an oak wood in March. The wood is still bare of leaves and so full of light. Warm spring days wake up hibernating insects which pollinate the flowers and so ensure their proliferation. Primroses are also found in hedgebanks and on western sea cliffs.

tree, often mixed with wych elm. Ash leaves provide a less impenetrable canopy; a great deal of light is able to filter through and, in consequence, these woods have a very rich ground flora. Ramsons or wild garlic, green hellebore, herb Paris, angular Solomon's-seal, giant bellflower, globeflower, wood geranium and the characteristic orchid of limestone woods, dark red helleborine – all of these and more will be found in the ash wood.

In commercially managed woodlands the best places for flowers are usually the rides, especially in high summer. If rides have been sprayed with weedkilling chemicals to keep them open they are very often barren, but fortunately this destructive practice is being superseded by a policy of alternate mowing: each side of a ride is mowed alternately every four years, and this allows a vigorous flower layer to grow. In high summer the rides will be full of foxgloves, St John's wort and rosebay willow-herb.

Woodland flowers have suffered badly from habitat destruction, and this is largely due to the modern demands for conifer plantations resulting in whole areas of broad-leaved woods being grubbed up and replanted with fast-growing conifers. Furthermore, old favourites like the primrose, snowdrop and bluebell were being extensively picked or transplanted to private gardens, although the public's increasing awareness of the need for conservation may have halted this trend. The age of a wood can often be gauged by its variety of spring flowers, and those with a particularly long list have probably evolved from the original ancient wildwood.

Woodland Birds

Bird population studies show that the number of pairs of breeding birds is higher in woodland than almost anywhere else, the average being some seven or eight pairs per 2½ acres (1ha), compared, for example, with only three pairs per 2½ acres (1ha) in general farmland. It is significant that these studies also show that most species do best in woods of over 250 acres (100ha) in size. Some species simply do not occur in smaller woods, and even nuthatches and woodpeckers prove to be most successful in larger woods, probably because food and territorial space are more easily available. Unfortunately most woodlands in Britain are now *less* than 250 acres, and this includes most woodland nature reserves.

The two most important requirements for any bird are food and a good nest site; oaks and sallows, which always support a large insect population will therefore be particularly rich in birdlife. A wood must also provide cover from predators, a reasonable freedom from human disturbance (which can usually be achieved by restricting people to paths and rides) and also 'song posts' – maybe a treetop, fence post or prominent branch – essential for proclaiming the bird's 'territory', and when attracting a mate.

Certain species prefer to feed and sometimes nest in particular trees – redpolls, for instance, prefer birch because the seeds form the bulk of their diet. Nevertheless, the species which usually comprise the bird population of a wood are surprisingly consistent, whether the wood be lowland or upland oak, ash or beech, alder thicket or ancient Scottish pine wood. Seventy per cent of the bird population is usually made up of the following eight species: robin, blackbird, wren, great and blue tits, chaffinch, and in summer, the willow warbler and chiffchaff.

An old wood with trees of mixed age – from forest giants to new-grown coppice stocks and shrubs – provides the greatest variety of food, nest sites, song posts and, therefore, bird species. Thus, the lowland pedunculate oak wood has a richer avifauna (birdlife) than the upland sessile oak wood: the climate of the latter is less favourable, there is hardly any shrub layer because the upland woods were managed as wood pasture, and since the soil is more acid, the field layer is less varied, particularly in the number of flower species. The level of insect life in upland oak woods is consequently lower, which means there are fewer birds than in the more bug-rich lowland oak wood. Birds needing dense thickets to nest in – blackcaps, blackbirds and whitethroats – are much less common in this environment. Furthermore, many of the trees in a sessile oak wood have narrow, deformed trunks in which the holes are comparatively small; these nesting sites are useless for woodpeckers (which need more space) but are perfectly adequate for pied flycatchers and redstarts which are far more characteristic of upland woods.

The most densely populated zone in many woods is the scrub layer, with more species breeding and usually in larger numbers than elsewhere. Its abundant insect life attracts not only many species of warbler and other insectivorous birds, but also tits, doves and jays. The high canopy is also rich in insect life, providing a further food source for these species. The larger birds, including rooks, carrion crows, herons and buzzards which occupy the canopy for nesting usually feed well away from the wood itself.

The seasonal cycle of the leaves has a dramatic effect on the variety of bird species and their life style; different groups of birds exploit different layers within the wood, and will move from one layer to another at different seasons. The main

The wren crams its dome-shaped nest into a variety of woodland cracks and crevices. This tiny plump bird is only 3¾in (95mm) long and has a distinctive short cocked tail. Extremely active, it forages among woodland litter, and catches insects in the vegetation.

event is spring, when huge numbers of seasonal visitors arrive, mostly insectivorous birds from Africa: willow warbler, blackcap, whitethroat, nightingale and pied and spotted flycatchers. These arrive to breed, their nestlings hatching in late May and June to coincide with the peak emergence time of millions of insect larvae.

By late summer the wood will be in dark full leaf and the resident tits, nuthatches and treecreepers will have grouped together, joining forces to search the canopy for the myriads of grubs and insect pupae. Many bird species of the woodland interior carry bright colour flashes so they can be

seen by other birds in the dark summer woods. The most well known of these is, perhaps the woodpecker. There are three species of woodpecker: the great spotted, the lesser spotted, and the green. The great spotted woodpecker can deliver some sixteen blows to a tree in under a second when drumming; the lesser spotted drums more softly for longer periods of two seconds or more. The great spotted woodpecker is about as large as a plump starling and may well now be the most common of the three species, found throughout Britain in suitable woodland. All three are absent from Ireland. The lesser spotted woodpecker is not nearly so common; it is about the size of a sparrow and is found in similar woodlands, especially old orchards. The larger, thrush-sized green woodpecker is found no further north than the Scottish lowlands and prefers open woodland and parkland. Its bright green and yellow plumage and brilliant red cap clearly distinguish it from its smaller black and white relatives. Population estimates are as follows: for the great spotted, 30,000 pairs; for the lesser spotted, only some 5,000 pairs; and for the green, 15,000 pairs.

The woodpecker is a classic example of a bird which has evolved to suit its environment. All woodpeckers excavate tree holes in which to rear their young, and besides a well-adapted beak and skull, woodpeckers have a remarkably well-developed foot and toe formation, enabling them to run about on a vertical tree trunk, with two sharp-clawed toes pointing forwards and two back. They also use their stiff triangular tail as a prop against the trunk. They can therefore search for food by clambering quickly up the trunk surface, hammering out grubs and other food with their chisel of a bill, and using their long sticky tongues to probe into holes and crevices.

Woodpeckers are seen most frequently in early spring as they display and drum in the woods, before the leaves are fully open. Later, their broken and striped colouring provides them with an excellent camouflage in the leaf-filled woods of high summer, and it is only with the return of winter when the branches are bare once more that one appreciates how striking these residents really are.

Holes provide secure nesting places for a variety of British woodland birds. The extremely rare wryneck is a member of the woodpecker family and was once found in many southern woodlands. Unfortunately it is now almost extinct as a breeding species in Britain, appearing only occasionally on the east coast in autumn, although it can still be found in the rest of Europe. It is about the size of a starling and, like its woodpecker cousins, it breeds in tree holes. A much more likely occupant of these holes, however, is the starling itself. Oak woods in particular are filled with starlings stuffing themselves with grubs in late June.

Most of the British titmice use tree holes for nesting. The familiar blue and great tits are the most common in woodland, followed by the less numerous tiny coal tit, although this usually prefers a patch of conifers. Marsh and willow tits are nearly identical in appearance and are both much rarer than the other three species. In the winter months they will all flock together to forage for food. Marsh and willow tits are easily distinguished by their quite different call notes, the marsh tit producing a loud 'pit-choo' note while the willow tit squeaks like a rusty hinge.

Whereas blue, great and coal tits use small holes and crevices for their nests and often choose unusual sites, marsh and willow tits prefer holes in soft, rotting tree trunks, usually quite near the ground. The willow tit in particular regularly excavates its own hole, tearing at the soft, rotten wood and flying off to scatter the chippings. It has strong neck muscles to perform this task, and consequently looks more 'bull-necked' than the slimmer marsh tit.

The old Scottish pine wood contains another hole-nesting tit: the rare crested tit; only found in the British Isles, in these last remnants of the ancient forest of Caledon.

Most young tits only stay in the nest for about twelve to fourteen days after hatching. By July, family parties have joined together to form autumn feeding flocks which may reach a hundred or more birds. Tit families tend to be large to compensate for the considerable losses which are suffered during winter, both to cold and to predators, and ten young tits jammed in a tiny hole in a tree stump is not unusual. A notable predator of tit nestlings is the great spotted woodpecker.

Estimates of the abundance of the various species of hole-nesting tits are as follows: blue tit, 3½ million pairs; great tit, 2 million; coal tit, 700,000; marsh and willow tits, about 100,000; and crested tit, under 1,000 pairs.

The nuthatch is a singular little bird, and has extra-strong claws to enable it to run as quickly down a tree as it runs up; in fact it has the three-forward, one-back arrangement of toes and so is more closely related to the tits than to the woodpeckers. Nuthatches are attractive birds and very noisy, despite their small size – their ringing cries echo through the woods throughout the spring. They use a woodpecker hole, but cement up the entrance with mud until it is small enough to pre-

Blue tits nest in small tree holes and are one of Britain's most numerous birds. Both the insect-rich scrub layer and the high canopy provide plenty of food. All members of the titmouse family can be found in woodland, though blue and great tits are the most common.

vent entry by larger birds such as starlings. They are common enough south of Yorkshire, but cannot be found further north, nor in Ireland.

Finally, the western sessile oak wood shelters two more hole-nesters: the lovely pied flycatcher and the fiery redstart, both summer visitors from Africa. The pied flycatcher takes well to nest boxes, and these have in fact been used for some forty years in an attempt to increase its population in many western counties. Pied flycatcher males are often bigamous; however, they may have one mate as far as 2 miles (3.5km) away from the other, presumably to encourage the first female to believe that they are unattached! The redstart can also be found in old woodland further towards the Midlands, but it is largely absent from eastern England.

Many bird species build their nests either high in the branches of the trees or in the thickets of the shrub layer. These nests are inevitably more exposed to the elements and to predators than those in holes and are usually expertly concealed or camouflaged. The large birds that nest in the upper branches of the bigger trees often obtain their food away from the wood – rooks and herons are the from early February onwards high in the treetops. Carrion crows also begin nesting early, but prefer to remain independent – their solitary nests can often be seen perched 50ft (15m) or more above the ground in the still-bare branches. Herons have been very accurately documented and there are some 9,000 pairs, with the largest colony of 200 pairs at Northward Hill RSPB reserve, in Kent. There may be as many as 860,000 pairs of rooks in Britain, but they are decreasing due to modern farming practice. The largest rookery is in Scotland, with a record of several thousand pairs.

At a lower level a variety of birds can be observed feeding and breeding among the middle branches – magpies, jays and wood pigeons are probably the most noticeable. The jay and the magpie are distinctive members of the crow fam-

ily; they are intelligent, but – in woodlands, at least – are shy and wary. Although it has been shown that the bulk of magpie food consists of grassland invertebrates, magpies and jays are also frequent predators of the eggs and nestlings of small woodland birds such as long-tailed tits, robins and goldcrests. If small birds build their nests early in the season these are usually less successful than later nests as they lack the leaf-cover necessary for concealment. In autumn the jay will collect and store acorns, its favourite winter diet. A jay's larder may contain several thousand acorns and any left uneaten will eventually help to regenerate the woodland.

The mistle-thrush also nests early, usually in March, and is another species which prefers to rely on a high, forked branch for a site, rather than hiding its nest at a lower level. Its loud song will be heard intermixed with the repetitive phrases of the song-thrush and the fluting call of the blackbird, both of which hide their nests in the thickets of the scrub layer.

The high branches may shelter a sparrowhawk, which builds its platform of twigs close against the trunk; from here it will hunt at high speed through the wood, snatching a small bird from its perch, taking it to a separate 'plucking' post – an old tree stump – to remove the feathers, then feeding the scraps of meat to its fluffy white chicks. The sparrowhawk is the wood's foremost avian predator, although many woods in the west and north may also support a pair of buzzards – larger, more powerful hawks whose basic design differs from the slim, swift sparrowhawk in that they have long, broad wings and a rounded tail, designed to quarter the ground, hovering occasionally, hunting rabbits and voles in the open.

The goshawk has recently started to colonise Britain, but is still very rare. It began to breed in Britain in the early 1960s and now the population of this shy forest dweller could be as many as 100 pairs; it may be that lost or released falconers' birds helped this increase. Goshawks are large, powerful and secretive; they are noticeably larger than sparrowhawks, and the female is much larger than the male, approaching the size of a buzzard.

The most densely populated part of the deciduous wood is the thicket, where the scrub layer meets the field layer of flowers, ferns and nettles,

although this lasts only for the short spring season from late March to June. Nearly all the insectivorous birds nest here from the resident robin to the closely related nightingale. In April most of the summer visitors arrive from Africa, although the first chiffchaff may sing from the treetops as early as March. All these birds rely on the new insect life in May and June to feed their chicks. Their nests are usually hidden at or near ground level in the dense vegetation.

The wren has an estimated population of 3–3½ million pairs. Male birds build several nests, known as 'cock's' nests – up to twelve by one male has been known – and the hen will then choose one to line for the eggs. In winter many wrens may roost together in tree holes or nest boxes, and as many as seventy-eight have been seen leaving one box.

The goldcrest, at 3½in (9cm) and ⅙oz (4.5g), is the smallest bird to be found in deciduous woodland. The wren is the same length, but fatter,

LEFT
A female sparrowhawk perches on her plucking post clutching her prey, a luckless sparrow. The female sparrowhawk is much larger than the male and her underparts are closely barred with grey, while the male's underparts are barred with reddish-brown. The sparrowhawk's flight normally consists of a few rapid wing-beats between long glides.

RIGHT
A cock chaffinch looks for seeds in a winter woodland hedge. Together with the blackbird, the chaffinch is one of Britain's two most common birds, breeding in woods, scrub, hedgerows, parks and gardens. The male's handsome plumage is distinctive; the female is less colourful, but shares the male's white shoulder-patch which is a useful identification mark.

weighing about ⅓oz (10g). Most small warblers like chiffchaffs and willow warblers weigh about ¼–½oz (7–12g) and yet they still make prodigious migrations of some 4,000 miles (6,500km) each way from Africa, and often return to the very same clump of trees!

By the time autumn arrives, as the summer birds depart and the wood loses its leaves, so the life-cycle of the remaining woodland birds begins another phase: from August until the spring of the following year the prime task is survival. They must find shelter and food. In late summer, warblers join the flocks of tits to fatten up on the insect store of the leaf canopy. As they migrate and the leaves start to fall in earnest, winter birds from the colder parts of Europe take their place: flocks of redwings and fieldfares, thrushes from Scandinavia, redpolls, siskins and bramblings, finches from the north – all these work their way through the woods in search of food. The permanent residents, the wrens, dunnocks, chaffinches and robins, scratch a living from the forest floor. The food source it provides is of enormous value, and many species only survive because of the insects and seeds hidden within it. Robins and wrens eat the insects, great tits and finches take the beech mast and other seeds, while thrushes eat the snails.

The bird population typical of conifer groves is strikingly different from that of a well-established deciduous wood. There are also far fewer birds, even in spring when bird numbers are at their greatest, and not nearly so many species are represented. This is partly because in mature plantations there is usually no shrub or field layer of plants to provide a variety of food, and also because the trees are felled for commercial timber before they become really ancient, so there are no tree holes to provide nest sites.

The ancient Scots pine forests of Scotland are quite different, however, and often carry a rich shrub and berry-bearing layer in the clearings and glades which occur between the widely spaced trees, and these forests support a relatively large and varied population. Ancient Scots pines and dead trunks are full of holes, providing nest sites for many species – the gem of these ancient Caledonian forests is the rare crested tit. Crossbill, siskin and goldcrest are all conifer woodland specialists, and, if there are a few birch trees they

may well be joined by redpolls which have expanded their range and population in Britain. Spring invariably heralds the arrival of a wave of summer residents from Africa such as willow and wood warbler, redstart, tree pipit and spotted flycatcher, all of which join the true conifer residents to breed in the scrub.

In a mature conifer plantation it is quite a different story, as only those birds which use the canopy have adapted successfully from living in an ancient woodland to living in these relatively recent man-made forests. The lack of holes has meant the disappearance of tits, with the exception of the coal tit which uses mouse holes, and the commonest birds are goldcrests with a scattering of siskins, crossbills and chaffinches. Crossbills nest early, many from early February, high in the conifers; they have one brood. Siskins also nest high in the conifer canopy; their population has increased in the north, with maybe 20,000 pairs in total.

These massive alien plantations, which have completely changed the aspect and habitat of many wild areas, pass through several stages of growth before their canopy closes over altogether. By this time, when the trees are perhaps 20ft (6m) high and fifteen years old, the plantation interior is dark, the shrubs have died down and disappeared and so too has much of the wildlife. However, in its two early stages, just after planting – the 'heath' and 'early thicket' stages – a conifer plantation is often very rich in birds, and mammals too.

The young trees grow with a wide assortment of heath plants such as heather, bilberry and bracken and the breeding birds at this stage include larks, pipits, curlew, grouse and many warblers. Because the plantations are fenced against rabbits, the field vole enjoys less competition and its populations increase; this in turn attracts the short-eared owl and the hen harrier, who take advantage of this abundance of voles to breed in these new and open forests. The short-eared owl nests on the ground; its population varies widely, from about 1,000 to 10,000 pairs in any year, depending on the cyclical population of field voles. Its young leave the nest at fourteen days and hide in the heather of the young plantation.

Once the trees are about ten years old the owls and harriers will have gone, since the timber

The coal tit is a bird of conifer woodlands, usually nesting in very tiny holes (including mouse holes) in banks and tree stumps. It is slightly smaller than the blue tit and is the only tit which has a white spot on the nape of its neck.

growth will have so re' .ced the ground cover, and consequently the population of voles. The warblers will be largely confined to the edges. Blackbirds, wood-pigeons and chaffinches appear at this 'late thicket stage', when the shrubs have been largely shaded out.

After fifteen years the trees are 'brashed', that is, all the lower branches are removed to reduce fire risk. In Britain these 'brashings' are usually left on the ground to prevent the growth of ground cover; in Ireland, however, bunches are tied to the trees to provide nest sites for wrens and thrushes. Brashing will drive most of the birds away at once,

leaving only the canopy nesters – jays, crows and magpies may use the highest branches, also predators such as buzzard, sparrowhawk and the rare goshawk. Sparrowhawks like a view from their nest and usually breed in the first few rows of trees, as do most of the small birds upon which they feed.

Many of these resident species of bird have physical characteristics which show their adaptation to their coniferous environment. Consider their feeding mechanisms, for example: the crossed mandibles of the crossbill are clearly adapted for prising out pine seeds. The larger of the two species, the Scottish crossbill, is capable of cracking tough Scots pine cones, while the smaller common, or European, crossbill has a lighter bill used on spruce. Similarly, the coal tit, crested tit and goldcrest have particularly fine bills to enable them to extract insects quickly and easily from between the pine needles.

Long-eared owls, though rarely seen, have apparently increased in the plantations since they do not have to compete with the more successful tawny owl of deciduous woods – they hunt out into the surrounding countryside. Long-eared owls are rare in deciduous woods on the mainland but in Ireland, where the tawny is absent, they occur widely. There are possibly 3,000 pairs at present. They use old nests, especially those of the magpie, high in the trees.

Although coal tits do occur in deciduous woods, it seems likely that their preference for conifers may be genetic. Experiments with hand-reared young blue and coal tits showed that the coal tits preferred pine branches whereas the blue tits used oak, just as they do most often in the wild.

Upland birds of the open moor as well as forest birds are threatened by mono-cultured alien forestry. More provision of nest boxes, especially on plantation edges, by forestry organisations would benefit many hole-nesters currently in danger. However, the most urgent requirement is conservation of the last scattered fragments of the wildlife-rich ancient Caledonian forests. Only a few square miles remain of a forest which once covered the whole of Scotland, and it is vital that no more of it should be destroyed.

Woodland Deer

Woodland deer have been one of the great wildlife success stories in these islands. There are deer in every county of Britain, and Britain is unique in Europe in having six species living wild in the countryside, although only two are truly native – the red and the roe deer. Fallow deer lived here between the Ice Ages, died out, and were reintroduced by the Normans. Deer in Britain are primarily forest animals, and those most likely to be seen in England, Wales and southern Scotland are fallow and roe. The red deer was originally a forest animal too, but was long ago driven into the hills – small isolated populations live on Exmoor, in the Lake District, the New Forest and the Brecks, but the majority are in Scotland, living on the open hills in summer and making their way down to the wooded valleys in search of food in winter.

Red and fallow deer have been kept in deer parks for centuries and these are very good places in which to become familiar with them. Sika deer, too, are kept in some parks. Sika originated in the Far East and several herds were brought over between 1860 and 1910 and introduced here. They are now found on the heathlands of Dorset and the New Forest, on the Yorkshire/Lancashire border and in many parts of the Scottish forests. There are well over a hundred deer parks in Britain, many of them private, and any deer you see, whatever the species, may be escapees. Many escaped during the two World Wars when fences were left unrepaired.

The tiny muntjac, also from Asia, has colonised large areas of the Home Counties and the Midlands, and was originally an escapee from Woburn Park. It is very shy and difficult to see, and is no taller than a fox.

You will know there are deer about from their tracks in muddy paths and from the presence of droppings. However, to see them you must be prepared to spend some time quietly waiting, preferably around dawn or dusk which are their most active periods.

Deer have superb senses of smell and hearing; they can detect your scent from as far away as half a mile (1km). One whiff of the scent of man and they will fade back into the landscape. So check the wind direction and approach *into* the wind. Find a vantage point or a forester's high seat overlooking the wood edge, or a clearing at a junction of

A watchful roe deer doe surveys the landscape with suspicion. The smaller of our two native deer, it is widespread in northern England and Scotland and also common in central southern England. Midway in size between the fallow and muntjac deer, it does not have spots and has much simpler antlers than either the red or the fallow deer which it sheds in November.

Find a vantage point or a forester's high seat overlooking the wood edge, or a clearing at a junction of two rides, and sit still. Then you must be patient, because if there *are* deer around they will almost certainly have noted your arrival.

Roe live in small family groups or even singly, so it is rare to see more than two or three together. Their rutting (breeding) season is in late July or early August, two months earlier than the larger species, the red, fallow and sika deer whose rutting season may start in September or October and last for several weeks. These three species live in herds for much of the year, and during the rut the largest red and sika stags and fallow bucks try and round up as many hinds or does as possible.

Red and sika deer shed their old antlers from spring onwards and new ones start to grow immediately, initially inside a protective coat of velvet. Once the antlers are fully grown this is rubbed off against branches, often causing damage to the trees. Roe grow their antlers through the winter and shed their velvet in spring. The antlers of fallow deer are cast in May and regrow by August. Gestation is usually about seven months, so young deer are born in early summer, roe usually in May and the rest in June. Muntjac are rather different; their antlers are shed in autumn and they appear to breed at any time of year.

Deer graze on the wood edge and browse the undergrowth. Park deer produce a noticeable browse line at about 7ft (2m) on trees within reach, and roe are particularly fond of bramble and roses in the lowlands and heather on the moors. Deer are often surprisingly noisy: as they graze they crack twigs and snort and bark, and during the rut the larger deer roar and groan very loudly. If a single deer is suddenly alarmed it will sometimes bark a sharp warning: sika, roe and in particular muntjac bark like dogs, and if you have been sitting quietly in the dusk, and a nearby, unseen deer barks loudly, you will probably jump out of your skin!

Fallow deer reach an average length of 5½ft (170cm) and a height at the shoulder of 3ft (92cm). The antlers appear in the buck's second year and reach full size by the sixth year; its average weight is 10½ stone (66kg). The average length of roe deer is 4ft (120cm), shoulder height 2–2½ft (65–73cm) and weight 3½–4½ stone (23–27kg). The largest species is the red deer which may grow to 30 stone (189kg), although the weight varies according to the locality. Average shoulder height is 4ft (130cm).

If you find a new-born deer, you should never touch it, despite its appeal. It has not been deserted, and its mother is hidden nearby, anxiously waiting for you to go!

Squirrels

Squirrels are the easiest to see of all the woodland mammals because they are active for much of the day. The true native British squirrel is the red squirrel, but the common species known throughout most of the country is the grey squirrel, introduced from North America between 1877 and 1920.

The grey squirrel has for the last forty years been regarded as a serious woodland pest, and has now spread to most of England and Wales and into the Scottish lowlands, largely replacing the native red. There are no red squirrels nearer to London than the small populations to be found in the East Anglian forests and on the Isle of Wight, and Brownsea Island in Poole Harbour. Recently a small number were fitted with radio collars and released in Regent's Park in London, to discover if the species could re-establish itself in competition with the resident greys; however, they have not apparently survived.

The native red squirrel – 'Squirrel Nutkin' of Beatrix Potter fame – has a much better public image than the grey, but it used to be a serious pest. It is primarily a conifer species dweller and appears to have been on the decline for many years before the grey even arrived, due both to deforestation and disease. The red squirrel now lives

Young grey squirrels, about four weeks old, view the outside world for the first time. They are born in dreys – untidy collections of twigs wedged in trees. Split nuts and gnawed pine cones often indicate a squirrel's presence.

In its native America the grey squirrel largely inhabits broad-leaved woodlands. It is heavier than the red, laying down more fat and feeding more on the ground, and is an altogether more adaptable animal, taking advantage of man's own environment and living in parks, gardens and hedgerows – anywhere with a few trees to hold its drey. The drey, or squirrel nest, is an untidy ball of twigs and leaves usually situated high in the upper branches; red squirrel dreys are smaller and more compact than those of the grey. Both species will also use tree holes.

In summer, most greys have a very definite reddish hue to their coat and many people will mistake them for a real red squirrel; this is just the lightweight summer coat which moults to thick winter grey in late autumn. Red squirrels are smaller and an orange-red colour and in winter have ear tufts which are quite unmistakeable. Both are extraordinarily agile, able to scramble up and down tree trunks with great ease, aided by long claws and truly amazing double-jointed feet. Both are remarkable tree-top acrobats, using their wide tails for balance and to steer.

There are usually signs of squirrels all about you as you walk through the forests. One tell-tale sign – apart from their dreys – which is especially noticeable in winter would be a number of gnawed pine cones and split nuts littered about the squirrels' favourite tree-stump 'tables'. Also, autumn fungi will show toothmarks and there will be footprints in mud, and snow, for squirrels do not hibernate. Stripped bark is yet another sign of their presence and this is their main crime against forestry; they will strip the bark off above ground level, especially in spring when nuts and seeds are scarce, searching for the sweet sappy material underneath. This can cause enormous damage to the tree.

The grey squirrel feeds on the ground much more frequently than the native red squirrel which it has largely replaced. Although it sometimes looks quite rufous, the grey squirrel can be distinguished by the fact that it never has ear-tufts. The grey squirrel has become a serious forestry pest, especially through its habit of stripping the bark off trees.

Surprisingly, squirrels find least food in summer, when there are no succulent new shoots and no nuts or seeds. Young squirrel mortality at this time is very high and few adults survive beyond two years, although squirrels have been known to live for five or six years. In winter, both species will be found out and about every day except in very cold or windy weather, greys searching out the stores they buried in autumn, using their memory and keen noses, reds searching the treetops for cones. Inevitably, some of the buried nuts and fruits will be missed, helping in the natural regeneration of the forests. Squirrels are rodents and therefore incessant nibblers, constantly needing to wear down their incisor teeth which are always growing; these great teeth are splendid nutcrackers and can split a hazelnut neatly in two.

Baby squirrels are born in two litters, one around March and the second from July onwards, taking up to three months to become weaned and even partially independent. May and September are thus good months to locate families of half-grown squirrels scampering at high speed over their home tree.

The average weight of the red squirrel is 10½oz (300g), and of the grey 1½lb (550g), around twice the weight of the red. The average length of the red squirrel to the tip of its tail is 1¼ft (40cm) and of the grey 1½ft (50cm). They have similar breeding patterns; newborn grey squirrels are nearly ten times as heavy as newborn reds. Squirrels can travel at 18mph (30kph) and leap 13ft (3.7m).

Badgers

Badgers have lived in Britain since the Ice Age, and can be found throughout almost the whole country, being most numerous in the south and west. However, in East Anglia they occur either rarely or not at all, and the same is true of most of Yorkshire and Lancashire, and of the Highlands and Islands of Scotland.

Badger setts are distinctive: most setts have about half a dozen large holes, and a spoil heap of soil thrown out of the front indicates a cleaned-out hole in current use – these are normally obvious,

though large setts may have many entrances. Some have as many as eighty to ninety, and may have been in existence for several centuries, stretching into the hillside for as far as 100yd (95m); the average, however, is 20yd (18m). Badger setts have been found at altitudes of up to 1,700ft (520m). A large sett may still have only one pair of badgers in residence. Fresh paw marks just inside the sett entrance are a good sign of occupation – badgers have a five-toed foot with all the pads pointing forward, and the footprints are quite distinctive. They are fastidiously clean, and dig their dung pits perhaps 11–16yd (10–15m) away from the sett; they also dig these pits at strategic places round the edge of their territory to help mark out their 'home' range.

Most badgers emerge at dusk, the boar usually appearing first, especially in spring when cubs are present. There are normally two or three cubs which are born in late February or March deep in the sett; they will appear for the first time in mid-April when they are about six weeks old. By July they will be nearly the size of their parents and fully weaned.

After a good scratch the boar will usually trot off down a path, and shortly afterwards the cubs and the sow will appear. Sometimes the badgers may stay around the sett for a while, especially if the cubs are small, when they will play and 'rough and tumble' in front of the entrances. A hardened, flattened spoil heap at the sett in May often betrays the presence of cubs.

Adult badgers are round and thickset, about the size of a cocker spaniel, and some 1½–2 stone (10–12kg) in weight. Their legs are short and powerful with strong claws, their fur rough and thick, the grey colouring the perfect camouflage for their twilight activities. The distinctive black and white face mask, so easily seen, is used as a contact signal

The badger's black-and-white facial mask functions as a recognition signal in the darkness. Commoner in wooded districts than generally supposed, the badger is present throughout the British and Irish mainland. Badger-watching is an increasingly popular pastime and many authorities take care to provide badger runs where new development crosses established badger paths.

but is probably most effective as a warning to potential enemies particularly for the cubs which, if attacked, fluff themselves out and face an aggressor head on. Badgers have huge jaw muscles attached to a distinct crest of bone on their skull, and anything which has faced an adult's bite will think twice before tackling a cub.

A badger's diet is truly omnivorous and includes anything from slugs to bluebell bulbs to wild honey; however, earthworms provide the bulk of their year-round food. One interesting trait is their habit of leaving little rolls of bedding material around the sett, usually dry grass or bracken – they will bring bedding into the sett quite frequently especially when the cubs are small in early spring, tucking it beneath their chin and dragging it in backwards down the hole.

Badgers happily tolerate man provided the immediate vicinity of their breeding sett is free of disturbance; some setts are active even in the heart of a town. Badger enthusiasts who observe these creatures from their own garden entice them successfully with food.

Unfortunately, many of these charming animals are killed each year on roads and railway lines, and this is the greatest single cause of mortality. At one time they were suspected of carrying Bovine TB, and the Ministry of Agriculture carried out a campaign of control by gassing. Thankfully this culling has now ceased, due in part to insufficient evidence that badgers were in fact the cause of the problem.

Bats

Bats may evoke all sorts of fearful reactions in humans, all stemming from our superstitious past. In fact most bats are tiny, furry, pleasing little animals, and scrupulously clean. There are some nine hundred species worldwide, but only fifteen are regularly recorded in Britain. None of Britain's bats is in any way harmful and all are small; the smallest is the pipistrelle, half the weight of a blue tit and only 1½in (4cm) long – it will fit into a matchbox! The largest is the very rare mouse-eared bat, a little larger than the noctule which is much more common. Even so, the mouse-eared bat is only 4in (10cm) long, not much bigger than a blue tit, with a long wing span.

Britain is on the edge of the Continent and cold by comparison with the more southerly European areas, so there are fewer species resident here; yet bats still make up a third of the mammal species present on these islands. All fifteen bat species are represented on the warm south coast; about ten in the Midlands; but only two or three in the far north.

The pressures of modern civilisation have caused bats to decline dramatically throughout the country: loss of food due to pesticides, loss of roost sites due to removal of ancient woodland, and loss of habitat because of widespread urbanisation. Modern timber treatments, the intensification of farming and the increased use of agricultural pesticides have all severely reduced the population of insects on which the bat feeds. Bats are much more difficult to assess and study than birds, the other main insect-eaters, and the best information comes from an estimate of numbers at winter hibernation sites; these have fallen dramatically, and some species have completely vanished from large areas. The greater horseshoe bat in particular has declined considerably.

Bats are mammals, the female usually producing only one tiny baby which is pink and blind, and which sucks milk in the way of all mammals. The young are often carried by the mother for the first two or three days, but are then 'hung up' to roost on a vertical surface – a wall, or inside a hollow tree; they will roost in this way throughout their life. Their eyes open on about the seventh day, and they take their first flight when about twenty-two days old. All bats in Britain are nocturnal and they are true flying mammals; their wings consist of completely airtight membranes stretched over elongated finger bones. They hibernate from late autumn to spring and are really only completely active for some six months of the year; even during this period they are torpid for some twenty hours each day. In hibernation they may use up 30 to 35 per cent of their body weight.

Most species of British bats are found in or near woodland and are easiest to see at dusk or dawn, especially along the wood edge or round lakes and ponds where they feed. They are almost impossible to identify unless held in the hand, and a licence from the Nature Conservancy Council is required to do this. However, the following four species are the most numerous and can be identified in flight with some certainty:

Pipistrelle – most numerous, very small, frequent twists and dives
Noctule – large, starling size, flies high and powerfully; sometimes dives low after insects
Long-eared – medium, broad wings, slow hovering flight close to vegetation
Daubenton's – medium, pale-coloured underside, flies quickly, skims surface of ponds.

Bats are not blind at all; indeed, they can see quite well. They do catch their food, however, in the dark, by a remarkably sophisticated system of echo-location, one which is still far in advance of modern man-made radar. They will emit about ten 'shouts' or squeaks per second of high-frequency sound, far above our own human range of hearing, and will then carefully examine all the echoes. This system is clearly incredibly sensitive, as a bat can even detect insects crawling on vegetation. Some large moths have evolved the ability to 'hear' these ultrasonic signals, giving them a chance to dodge an oncoming bat.

Finding out exactly what constitutes a bat diet is very difficult and most has been learned by the microscopic analysis of dry droppings taken from roost sites. They clearly eat large quantities of insects – individual bats have been known to gain 25 per cent of their own bodyweight after one night's feeding, which means a single pipistrelle may eat 1,000 gnats a night! A roost of 100 pipistrelles

The pipistrelle is the smallest and most numerous of Britain's fifteen species of bat. It can be seen in many different kinds of habitat and frequently over water, where it feeds on flying insects. It can normally be seen about twenty minutes after sunset; it is rarely seen in the daytime.

would therefore easily clear a million gnats a month and many other annoying or harmful insects are included in their diet. We should consider this a useful service, yet we have harassed them in places to vanishing point, especially in roosts in houses, and this is really because of nothing better than ignorant prejudice. Most species roost in hollow trees in summer, but pipistrelles favour the roof-tile spaces in modern houses; some others use lofts and some breed in caves.

Bats build up their fat reserves in autumn and then hibernate quietly throughout the winter in cave crevices, behind tiles on houses and in hollow trees, the choice depending on the species, waking occasionally to drink and perhaps locate food if it is not too cold. Hibernating bats should not be disturbed, as this may drain their fat reserves too low to survive the rest of the winter. By spring all their fat reserves will have gone, and then they must feed or starve. Cold springs can be disastrous for already depleted bat populations.

It is clear that bats still need friends despite the protection afforded by the 1981 Wildlife and Countryside Act. There are two particularly significant ways in which they can be helped: firstly, by treating roof timbers with permethrin instead of lindane, which is lethal to them; and secondly, by not spraying the loft while the bats are present – woodworm will not be particularly worse for waiting a while, and the bats will usually move on in a few weeks. House owners who are determined to get rid of their bats are legally obliged to tell the Nature Conservancy Council, who will advise on how to displace them without killing them. Fortunately most people are more conservation-minded now, and actually enjoy the spectacle of their bats emerging at dusk.

43

Small Woodland Mammals

The small mammals of woodlands, the mice and voles, are often very numerous, and the wood mouse, or long-tailed field mouse is the most widely distributed of all. The four layers of vegetation in a deciduous wood are all occupied by mammals, although the upper branches of the canopy would normally only shelter the two squirrels, the dormouse and certain species of bat roosting in tree holes. Lower down, the shrub layer provides food for several species of small mammal which are all capable climbers: the most frequent are the wood mouse and the bank vole, but the true champion of this particular community is the much rarer dormouse. The common or hazel dormouse has in fact declined dramatically over the last fifty years or so, primarily because of the destruction of old woodlands. Another dormouse which is even rarer is the fat dormouse, found in a small area of the Chilterns as a result of an introduction at Woburn Park in 1902.

Most small woodland mammals live down on the forest floor and are difficult to see because they are usually nocturnal, although the bank vole is diurnal in habit. In order to observe them it is necessary to put out some bait – some peanuts or a few sultanas – on a clear mossy bank. Use a torch – a small one at first, then graduating to a more powerful one over two or three days – with a red plastic filter over the front to cut down the glare. Most nocturnal mammals are far less sensitive to red than to white light.

The wood mouse will probably be the first visitor, and it is quite different from the grey house mouse. Its larger eyes and ears enable it to cope more effectively with its nocturnal lifestyle, and its russet-brown top coat and white underside camouflage it better. Close views will show a little patch of yellow fur on the throat. Young wood mice appear from May onwards, and are often grey in colour.

The wood mouse is common all over the country, though less so in the mountains. However, it has occasionally been found on the high tops, even on Snowdon summit where one was found living on the scraps left by tourists. It is found in nearly all types of habitat including sand dunes, but woodland is still its preferred home. It appears to dislike open grass fields which is just as well as it therefore avoids having to compete with the field vole for food. Field voles may be found in the hedge or along the wood edge bordering grasslands, but very seldom move further into the wood.

Wood mice develop an extensive run system in the leaf litter but will forage for food anywhere within their home range; they may even climb into the branches after fruit and insects. However, they are largely seed eaters and store nuts and seeds in their holes. The 'home range' is dictated by food supply and density of population, and may be no more than 20yd (18m) square. Wood mice are nocturnal, with peak activity at late dusk, and just before dawn in winter, and a single midnight peak in summer. Activity can be inhibited by bright moonlight.

Wood mouse population is at its highest in autumn when all the adults and young of that particular year are still alive – and there may be six or seven litters! Each litter contains five or six young, which are weaned at three weeks. Wood mouse numbers seem to remain high well into winter, but then there is a rapid decline in very early spring, often the coldest and most hungry months, until a lower, more stable level is reached by the start of the following year's breeding period in April. The young then have a better chance of survival as the older members of the community die off.

The yellow-necked mouse is true to its name and has a conspicuous yellow collar; at just over 1oz (33g) it is two-thirds larger again than the wood mouse. It has slightly different population dynamics (probably in order to avoid feeding in competition with other mice), its numbers increasing steadily in spring and summer. The winter survival rate, however, is low. Survival of all the woodland small mammals is wholly dependent upon the seed and berry crop which varies greatly from year to year; population density varies accordingly.

Small mammals seem to be more dependent upon the structure of a wood than any particular plant or tree. Where there is an open field layer, the wood mouse is the most numerous rodent, while the bank vole predominates where there are dense herbaceous areas. Bank voles are easily recognisable, with blunter faces than mice, their ears short and flattened along the head, their fur deep chestnut above and cream or grey beneath. They feed on berries and leaves, while the mice take the harder nuts and seeds. Like many small mammals, bank voles produce several litters during the season, and the young may be sexually mature within five weeks.

If you are lucky you may see a hazel dormouse but it is a rare creature nowadays and strictly nocturnal, although perhaps as a result of this it has to a certain extent been overlooked. It lives in ancient woodland where there are many layers of vegetation so it can climb quickly through the branches, both for food and to escape predators. It is now known to eat a wide range of nuts and berries. Its preferred habitat is hazel coppice filled with honeysuckle, where it builds its summer breeding nest of stripped honeysuckle stems and grass, usually situated above ground level but in dense vegeta-

tion. There are probably only two litters a year, and sometimes only one, with three to four babies which take some five to six weeks to become independent.

Unlike the other small mammals the dormouse hibernates from late October right through the winter until April. It is therefore best seen when its population is at its highest in late summer or early autumn and as it searches the shrubs for food, fattening itself up to survive the long winter months. The best way is to walk quietly through the woods at dark with a powerful torch – although it is possible to search in this way for days, even in a wood known to contain dormice, without seeing one at all. Its weight just before hibernation when it is at its heaviest is around 1½oz (45g).

Fortunately dormice take readily to bird nest boxes, especially if these are baited with a little chopped apple. However, it must be appreciated that it is illegal even to handle a dormouse without a Nature Conservancy Council licence, because of its rarity.

A wood mouse suckles its young in an underground nest chamber. Largely nocturnal, it has large ears and eyes to compensate and is the commonest countryside mouse. Its diet consists mainly of seeds, fruit and insects.

Woodland Butterflies and Moths

Butterflies need sunshine, and their caterpillars need food plants, so the woodlands which can offer many clearings, glades and rides suit them best. Because they need a high degree of warmth and sunshine to thrive, the south of Britain has a larger number of resident species than the north, but as far north as the Scottish Lowlands butterflies are an obvious and important part of the woodland fauna. They and their larvae form an important link in a complex food chain, supplying the small insectivorous birds which in turn feed the predators at the top – the hawks and foxes, sometimes even magpies.

Butterfly numbers and the variety of species are largely dependent upon sensitive woodland management, but some of the best woods may hold as many as thirty-five species during the year. Forty of the sixty most common British species can be found in a woodland habitat, representing six of the eight families. The woods which provide the optimum environment for butterflies, and for most other insects too, are those with many layers and a variety of plants, for example with high oaks and ash at the top, then with clumps of thorn and sallows at a lower level, and finally at field level, rides and clearings full of flowers.

The age-old practice of coppicing provides clearings at regular intervals and happily has been revived in many woodland nature reserves. In unmanaged ancient woodland, fallen trees would let in life-giving sunshine, resulting in open glades; nowadays man himself provides broad rides in order to extract timber. Many commercial woodlands have rides simply to divide the forest into more manageable blocks, but provided these are mown late in summer after flowering they are often full of butterflies. Rides which are cut too early or – worse still – sprayed with chemical herbicides to keep them open are usually barren.

The first warm sunny days of early spring will encourage certain species to emerge – namely those which have hibernated in adult insect form. Bright yellow brimstones are the most obvious, and the occasional tortoiseshell or peacock, small and colourful, fluttering up from a sunny spot on the ground where they have been soaking up the sun.

To improve one's chances of seeing different butterflies it is useful to know when the individual species are likely to emerge, as well as something about their habits and the food plants required by their larvae. Some sixty species are native to Britain, while twenty more are rare migrants from abroad. The yellow brimstone is the first to appear – some speculate that its bright yellow colour gave rise to the 'butter' in 'butterfly' – in early spring. The first to emerge from overwintering pupae are the speckled wood – found in dappled glades in April – and the orange-tip, both of which appear together with the main mass of spring flowers: the violets, bugle, anemones and bluebells of a British wood.

Butterflies seek out their food using scent and vision. Their compound eyes are made up of some six thousand 'ommatidia' or single-lensed eyes. They can see colour clearly; many are highly coloured for the purpose of differentiation and some are sexually dimorphic, that is, the sexes are different colours. Furthermore they can see well into the ultraviolet end of the spectrum, and many flowers have evolved an ultraviolet pattern, not visible to humans, which aids pollination by attracting the butterfly (or other insect) to the nectaries.

When a butterfly takes to the air for the first time it starts with a full set of wing scales, maybe two million or more. High speed cine-film has shown that showers of scales are shed at the flap of each wing.

Courtship begins soon after the butterfly emerges from its pupa, the male being attracted to the female by her strong scent emissions. Male

The purple emperor butterfly is rare but widely distributed in southern and central England, principally in ancient oakwoods. Its caterpillars feed on sallow.

Its powerful flight and speed coupled with its brilliant iridescent wings make it a breathtaking sight. The purple colouring is only present in the male.

butterflies have modified scales on their wings which produce scented pheromones; these are 'showered' on the female during courtship flights, and help induce mating.

A rush of emerging butterflies appears in late spring, benefiting from the mass of nectar-filled flowers. Colourful commas, fresh from hibernation, grizzled and dingy skippers along with common blues, whites and green hairstreaks – all feed busily along the rides. As spring changes to summer another group of larger, brightly coloured butterflies appears, the fritillaries, browns, white admiral and the king of the woods, the uncommon purple emperor. Many of these species appear in late June or July when most woodland flowers have finished flowering. Glade and ride flowers are then particularly important with tall marsh thistles and bramble blossom providing much-needed food. Despite their size and bright colours, most have superbly camouflaged undersides, allowing them to sit with their wings closed, concealed from predators among the vegetation. The wood will therefore seem quite deserted if on a dull day the butterflies do not appear.

Despite the substantial list of woodland butterflies only one, the purple hairstreak, actually feeds on a tree, the oak. Its larvae eat the oak buds as they burst into leaf and, like the larvae of most species, are camouflaged to look like their surroundings.

High summer brings a new brood of brimstones, peacocks and tortoiseshells which fly with the other woodland butterflies before hibernating until the following year when the cycle will start all over again. Most other species fly for two or three weeks, and then die after laying their eggs. Some species, such as the hairstreaks, the chalkhill and silver-studded blues, the Essex and silver-spotted skippers, pass the winter as eggs. Most survive as caterpillars, which can cope with 5°F (−15°C) of frost because they have a natural 'antifreeze' in

A brimstone butterfly emerges from hibernation on the first warm day of spring. Feeding only on alder buckthorn and purging buckthorn, it is found in wooded or open country. The bright yellow colouring is not present in the female but, with wings closed, it is sometimes difficult to tell the male and female apart.

their body fluids. Warm weather enables caterpillars to develop more quickly; furthermore, they can accurately measure the difference between fourteen and a half and fifteen hours of daylight in autumn, at least in the case of the large white. If there are less than fifteen hours the caterpillar will change into a chrysalis and stay like this until the following spring. If more than fifteen hours, then the chrysalis produces a butterfly a week or so later.

Unfortunately butterflies have been in dramatic decline over the last thirty years, largely because of man's use – or more often abuse – of the countryside.

Moths evoke far less general interest in people than do butterflies despite the much larger number of species: there are over two thousand species of moth in Britain against only sixty resident butterfly species. The butterflies are more noticeable, of course, since they are all daytime insects while most moths are nocturnal, and often a little drab in colour. There are, however, many colourful moths and many with strange life-histories.

British moths are subdivided somewhat arbitrarily into two groups, the macro- and micro-moths. There are some twelve hundred species of micro-moth, many tiny enough to require a magnifying glass to examine them. At the other extreme are the large, fat-bodied hawk-moths, some with a 5in (13cm) wing-span. Moths are found in most of Britain's habitats, but woodlands provide the optimum conditions and the best variety of

The red underwing is a large moth with a 3in (7cm) wing span. It lays its eggs on sallow and willow leaves on which the caterpillars feed. The adult moth flies in August and September and is found mainly in the south and east of England.

food for the larvae. Several hundred species can be found in a mature wood, especially if it is ancient woodland evolved from the original post-glacial forests with the concomitant wide variety of plants.

Nearly every single different plant species will have at least one moth larva specifically adapted to feed from it. Some, such as sallows and willows, may harbour well over 100 species, including the eyed and poplar hawks, the puss-moth, several prominents, red underwing and the herald; while the oak plays host to up to 200 different moth larvae, including the lobster, buff-tip, vapourer, merveille de jour, copper underwing, lackey and oak-eggar moths.

Most moths lay their eggs in autumn or early winter, although some species can survive the coldest days. Most of these eggs hatch in the late spring to take advantage of the breaking leaf-buds, exploiting the soft foliage of the young leaves. Moth caterpillars are a vital part of the ecology of woodland, providing food for nearly all the birds in these areas. Migrants from Africa return for their short early-summer breeding season in time to take advantage of the mass hatching of millions of caterpillars; and in May and June the fresh green oak crowns also teem with starlings, tits and even rooks and jays which have changed their usual diet to gorge on the sudden glut of food.

Sometimes one species of moth may reach plague proportions and then a tree or even whole areas of woodland can be completely defoliated. The oak tortrix moth is only the size of a fingernail and looks quite harmless. Yet if undisturbed by the birds for whom it normally provides abundant food, a colony can strip a tree in a matter of days – clouds of little pale green moths will flutter down like snowflakes from the oak branches, with as

many as 300,000 in one tree.

Camouflage is essential to avoid being eaten, and to this end moth larvae display an amazing variety of colours and forms. Both larvae and adult moths adopt many weird colour patterns and disguises to outwit their predators. The caterpillars of the geometer family of moths are called 'loopers' and usually so closely resemble the twigs of their food plant that they are only visible if they move. They can remain quite motionless, anchored to the foliage by only their hind claspers. Many moths, such as the eyed hawk-moth, rely on the intricate colour pattern on their wings to disguise their outline while they are at rest. Some caterpillars are disguised to predators because they appear to be something quite mundane: the black hairstreak pupa, for instance, looks like a bird dropping! Others set out to appear fierce and aggressive – the elephant hawk-moth larva has large eye spots which it can suddenly expose to frighten a predator.

Sometimes taste is the deterrent. For example, the death's-head hawk-moth and the cinnabar moth both have conspicuous caterpillars which eat plants containing strong toxins. These pass into the caterpillar and make it unpleasant to eat, sometimes even poisonous. Predators soon learn to leave such caterpillars well alone. They are often very striking with bright bands of colour which act as a warning to would-be enemies.

Some moths lay their eggs deep in old wood, the most spectacular of which, the goat-moth, produces larvae which grow to 4 or 5in (10–13cm) long; these take several years to mature and when they finally emerge they leave an exit hole in the timber 1/2in (1cm) wide.

There are various ways of obtaining moth specimens undamaged for closer examination. The method of collection probably most often used by the local field studies centres is a mercury vapour trap; this is switched on at night in order to catch and identify a representative number of moths, and assess the population of the area. Moths are attracted by the ultraviolet light emitted, and tumble into the box beneath the lamp where they will sit quietly among the old egg boxes which are put there to give more perching space. With luck there may be several hundred to identify the next day, before they are released.

51

Woodland Invertebrates

The variety and number of insect species and the host of other invertebrate life forms to be found in woodlands occur in bewildering array. Most of these creatures belong to the largest and most successful group in the animal world, the arthropods. This group includes all insects; also centipedes, millipedes and springtails; arachnids (spiders), ticks and mites; and the Crustacea, of which woodlice are the most obvious woodland inhabitants. All the arthropods have an external skeleton – 'exoskeleton' or 'cuticle' – and most of the adaptations to be found in this group stem from this one characteristic. Most of the remainder of woodland invertebrates are molluscs (mostly slugs and snails), Annelida (earthworms), Nematoda (roundworms) and Platyhelminthes (flatworms).

The most noticeable invertebrates are the larger insects. Apart from the butterflies and moths already described, there are hoverflies, bees, wasps, beetles and bugs. Craneflies, sawflies, gnats and midges are less noticeable, but are present in large numbers. Almost every source of organic matter houses some type of invertebrate, from deep down in the soil to the top of the highest tree. Some populations can be extraordinarily numerous; a large oak, for example, may support 3–400,000 caterpillars of the winter moth alone.

The largest number of creatures will generally be seen in the warmer months, as most overwinter in an inactive stage of development, as eggs or pupae; however, much of the leaf-litter fauna is more active during the cold, damp conditions of winter. Liniphyid spiders, for example, appear in their millions in late autumn and winter, even in upland woods and heaths; and in the hottest weather many species which need damp conditions aestivate, or bury themselves deep in the soil.

The variety of insects altogether represented in Great Britain is enormous, and many of these will be found in woodlands. There are thirty-eight species of grasshopper and cricket, all vegetarian – the long-horned grasshopper, in particular, may be found on trees and shrubs. There are nine species of earwig, usually found sheltering under bark or stones, and subsisting on a varied diet from scavenging. Stoneflies (thirty-two species) and dragonflies (forty-two) are nearly always associated with water, especially during their nymphal stages. The adult stonefly is vegetarian, all the others are carnivores; dragonflies will often be seen hunting along woodland paths at certain times of day.

True bugs are of the order Hemiptera, and there are some 1,630 species in Britain, including shield bugs, squashbugs, assassin bugs, capsid bugs, froghoppers and aphids. All have two pairs of wings, and mouthparts designed to pierce and suck. Shield bugs have a shield-like top plate, and are both plant eaters and carnivores. Capsid bugs comprise a large group including many which feed by sucking plant juices; for example *Plesicoris rugicollis*, feeding on willow, sallow and apple. Aphids and scale insects, usually present in vast hordes, feed by sucking plant juices, as does the large group of leaf hoppers, often on leaves of nettles, rose, oak and beech. Froghoppers are small jumping forms, the nymphs producing small blobs of a froth-like substance which can be seen on plants and is known as 'cuckoo spit'; *Philaenus* sp is a common example. Fifty-four species of lacewings lay stalked eggs which they attach to appropriately placed leaves, and their larvae feed on aphids. Certain species of alder and snake flies have similar characteristics: for example, two species of alder fly lay their eggs on vegetation next to water; and four species of snake fly inhabit summer woodlands and feed mainly on aphids.

Beetles are often strikingly beautiful, with some 3,700 species among the twigs and leaves. Many in their larval stage feed on leaves so are specific to particular trees; for example, the beech weevil and hazel weevil. Beetles have two pairs of wings, the hind membranous wings usually covered by the hard anterior pair. Beetle diets are many and varied: bark beetles – like the elm bark beetle which spreads the fungus of Dutch elm disease – excavate channels in wood beneath the bark. Longicorn beetles have long antennae and as adults feed on pollen; however, the larvae feed by boring into wood – the musk beetle larva, for example, bores into willow. Weevils can be identified by their projecting snout and are destructive plant feeders, the

A bumblebee sucks nectar from the flowers of yellow archangel. One of our commonest bumblebees, the buff-tailed bumblebee builds its nest underground sometimes using an old burrow or hole made by a mouse or vole. Its comparatively large body weight needs broad wings to produce sufficient lift to fly, and the thoracic muscles need to be warm to generate full power – hence the thick coat of body hair which cuts down heat loss.

nut weevil attacking hazel, the gorse weevil laying eggs in gorse seed pods.

Carnivorous beetles include ladybirds which feed on aphids, cardinal beetles which eat bark beetles, rove and ground beetles which feed on a variety of other insects, and scavenging species including dor and burying beetles.

The insect order Hymenoptera encompasses some six thousand species of bees, wasps, sawflies and ichneumons, all with two pairs of membranous wings, and all except sawflies having 'waisted' bodies. Ichneumons parasitise the eggs and larvae of other insects; thus the small apanteles wasp lays its eggs in the body of the cabbage white caterpillar. Gall wasps cause galls; oak apples, oak marble galls and oak spangle galls arise from three different species. Sawfly larvae are serious defoliators which can be distinguished from moth caterpillars because they possess pro-legs on the second abdominal segment.

There are 5,200 species of the order Diptera, or true flies, in Britain. These have one pair of wings, and most have mouthparts adapted for sucking fluids, either nectar or the liquid products of fermentation or decay. Some, such as gnats and horseflies, have mouthparts to pierce animal bodies. Carnivorous fly forms include the hoverflies, which are particularly remarkable because of their amazing powers of flight, being able to hover for long periods. They are superb in closeup, resembling bees and wasps, and possessing huge compound eyes.

The forest floor supports two groups of wingless insects: these include some 23 species of bristletail and silverfish, and 260 species of springtail, feeding on rotting wood and leaves. The forest floor provides yet another ecological stratum, for the many creatures which need the damp conditions of deep leaf litter; otherwise they would dry out completely and die. Some are vegetarian, and especially useful in recycling the dead material of trees and plants, returning its nutrients to the woodland system. Worms, mites, slugs, snails, millipedes and woodlice all contribute in this way; while spiders, centipedes and harvestmen are carnivores, preying on the vegetarian hordes.

Ferns, Mosses, Lichens and Fungi

The damp, shady floor of a wood provides the ideal habitat for many ferns, mosses and lichens. These are the non-flowering plants of a more primitive, 'lower' order, and many are of striking beauty. Altogether there are some ten thousand species of fern throughout the world; Britain supports only about fifty. It is their ancient lineage, however, which makes ferns so fascinating; they are a visible example of one stage in the past development of the flower, representing a link in the evolutionary chain concerning plants which has led to the extraordinary variety of flowers that we see today.

The shaded woodland environment provides the best conditions for ferns and therefore ferns uncurl their bright green leaves late in spring, and develop full leaf at a later stage in summer than most flowers; they then remain green far into winter before dying back to brown dead fronds. This cushion of dead leaves protects the living crown of the plant through the coldest periods of the winter.

The fern has no flowers, so it does not produce seeds: how, therefore, does it reproduce? The process of reproduction for the fern involves the use of spores which are carried on the fronds. Running all along the underside of a mature leaf there are tiny brown spore cases called 'sori' – the shape of which can help distinguish species. These spore cases continue to develop until the autumn, by which time they appear as dark brown nodules, full of spores: the cases will then suddenly split and the spores float away on the breeze – literally millions are produced by each plant. Such huge numbers are required because only a few will find the right conditions to grow.

The spores produced by the fern are asexual, unlike true seeds, but they will germinate in damp conditions and produce a 'prothallus', a tiny ribbon of cells which take some ten to twelve weeks to reach only a few millimetres in length; this must remain moist, and it must be covered with a film of water. Both male and female sex organs grow on the prothallus and when ripe they produce spermatozoa and egg cells; the film of water is needed so that the male sperm may swim over the surface to fertilise the egg cell – it is from this sexual fusion of cells that the new fern will grow. Thus ferns have both an asexual and a sexual stage to their reproductive process, and this is known as 'alternate generation' reproduction; furthermore it is this re-

A male fern unfolding its new spring leaves in April. Bracken prefers dry, acid soils and is the commonest British fern, found in woods and grassland. It is very invasive and can dominate huge areas.

Mosses and lichens growing on an old damp log. The spore capsules of the mosses are ripe and ready to burst in early summer. A minute lid at the top of the capsules springs off to show a ring of teeth which dry out and curl back, allowing the spores to be blown out by the wind. In wet weather, however, the teeth reabsorb moisture, straighten and so shut the capsule.

productive process which constitutes the essential difference between ferns and the more highly evolved seed-bearing flowers.

Bracken is the prime exception to this reproductive method of ferns: bracken in fact spreads itself by underground rhizomes, and this enables it to grow in much drier areas. But even bracken needs spores to spread to new areas, so it too bears elongated sori beneath its fronds.

A few species of fern – notably the hard fern, and the large rare royal fern – produce 'fertile fronds', and these are quite different from the other leaves. These carry the spores, and may be a living example of the evolutionary experiment which some 180 million years ago produced the real flowering plants.

Ferns can be found throughout Britain but are at their best in the damper conditions typical of the woodlands which grow in the Atlantic coastal regions. The most common species are the male fern and buckler fern. They are easy to recognise and quite distinctive. Ferns are of the order Pteridophyta, as are the horsetails, clubmosses and quillworts, all closely related and all producing spores.

Horsetails grow a small cone-like structure at the stem tip, and the spores produce an underground prothallus. There is only one group now, Equisetum, but millions of years ago there were huge forests of giant horsetails; these gradually died out and rotted down until ultimately they became a stratum of the earth. Much of the world's coal is a product of those original giant horsetail forests.

Clubmosses are few in number in Britain, and belong to an ancient group, the Lycopsida. They are mainly to be seen as small creeping upland plants, with spores produced in spore cases in the leaf axils. Quillworts are of the same group; they grow in water, and have round, grass-like leaves.

Mosses and liverworts belong to the order Bryophta, and there may be at least one thousand species of this group in Britain. They are related to ferns and have a similar life-cycle, and similarly, need damp conditions for reproduction. Like ferns, they are found in greatest profusion in the counties nearer the Atlantic but they are more widely spread than ferns. There may be thirty species of bog mosses alone in upland peat bogs. Mosses can be identified by their leaves which are either in spiral form, or sometimes in two flat rows; liverworts have either flat-lobed structures or small leaves in rows of three.

Mosses grow most rapidly in winter in Britain's cool, moist climate, and are most luxuriant in spring, producing spore capsules which burst in early summer to produce the sexual stages; these in turn develop into the asexual form that we recognise as the moss plant. Mosses in fact do not have true leaves, although the green filaments look like leaves; nor do they have any real roots. Generally, they absorb moisture over the whole of their surface, although a few can also take in moisture from below; they are therefore especially vulnerable to 'acid rain'.

Lichens are quite different from mosses and ferns, although they usually grow best in similar areas, and are unique in being a combination of an alga and a fungus. They can survive in much drier situations than ferns and mosses, the fungus penetrating the base material and the alga using sunlight to manufacture food. This means they can grow where no other plant can, and they are often the first to colonise a barren land. Since they do not reproduce by the alternate generation method, they are less dependent on water. The fungus provides the bulk and shape of the lichen with the alga occurring in a thin layer just beneath the fungal 'skin'. It is estimated that there are over one thousand species of lichen in Britain

Fungi are of vital importance in the natural world, and play a large and essential role in the rhythm of life. Many species live on dead material thereby preventing any excessive build-up of waste matter, breaking it down into simpler chemicals which can be recycled in the earth. They contain no chlorophyll and always obtain their food by breaking down other complex materials into simpler chemicals – the opposite of green plants which photosynthesise simple materials into more complex foods. The fungi that live on dead material – leaf litter or dead wood – are called saprophytes. Those that feed on living trees or animals can cause much damage: these we know as parasites.

In the human world fungi have many uses, and can exert considerable influence on our life-style, and even in our general well-being: they may be eaten; they are used to produce powerful drugs like penicillin; they help the production of such foods as bread, beer and wines that use yeasts in their manufacture. They may destroy our houses with dry rot; they may make us ill, or even destroy us altogether: fungi can therefore be said to affect mankind at all levels of existence.

There are several thousand species of fungi altogether in Britain, many microscopic, although some three thousand species have quite large fruiting bodies. The toadstools, mushrooms, puffballs, earth stars and other colourful shapes you see in autumn are merely the fruiting bodies of the fungus itself, which is hidden in the base material and is working throughout the year: the true fungus consists of a fine network of threads, the mycelium. The cap of the fungus has one prime function: to ripen and distribute the spores by

The magpie inkcap fungus grows in beech woods in late autumn. Fungi discharge astronomic quantities of spores: an ordinary field mushroom releases 100 million spores an hour from a total of 16,000 million; a giant puff-ball may produce 7 million million.

which the fungus reproduces. When ripe, the caps produce spores by the million, and huge numbers float away on the wind; they are produced in such vast quantity because only one or two will land on a suitable spot for germination. Most fungi rely on air movements to spread the spores, but earth stars, for instance, need splashes of rain, and stinkhorns rely on flies.

The larger fungi are divided into two groups, depending on the construction of the spore-forming organs. The Basidiomycetes, or 'spore droppers', form spores on the outside of a series of special club-shaped cells, the basidia. As the spores mature, they fall and are distributed by the wind. Most of the larger fungi are of this type, including the agarics, the boletes, the polypores and the jelly fungi. The second group, the Ascomycetes, produces spores within a club-shaped sac, (asci), and when these are sufficiently mature they shoot out through the tip of the sac. This group includes the morels, cup fungi and truffles.

Many fungi occur in a particular type of woodland environment; many are even found associated with one tree species only. For example, the familiar fly agaric, with its bright red cap and white spots, is usually found in birch woods, while the somewhat similar red-capped 'sickener' – so-called because in the past it was used as a powerful emetic – is found in pine woods. Others live directly on the logs or stumps of one tree species, for example, the fawn-coloured and translucent 'Jew's ear' is found on elder, while the white beech tuft grows on beech stumps. Some species such as honey fungus grow on living trees, and cause rot in the heartwood of the tree, destroying its valuable timber.

Other well-known fungi include the massive beefsteak fungus, usually found in autumn on oak and chestnut; the very pretty, banded bracket fungus which may be found on dead wood at any time of the year; and the striking stag's-horn, a yellow branched fungus which forms on old wood. Some fungi grow in great clumps; perhaps the most common of these is sulphur-tuft, another bright yellow, capped species which is often to be seen covering old tree stumps in October.

Most fungi grow best in damp conditions and often surprisingly quickly – a 4in (10cm) toadstool and cap can grow from a minute bud in the leaf litter in two days. Because the northern counties tend to have damp conditions earlier in the autumn, fungi appear up to a month earlier in the north than in the south (quite the opposite of green plant behaviour in the spring). Most of the fruiting bodies soon die down with the onset of cold weather, or are eaten by slugs, squirrels and mice. The mycelium, however, lives on below the surface – some species, particularly bracket fungi, live for many years, slowly growing in size.

The well-known phenomenon of the fairy ring can be created by several species, although the fairy-ring champignon, a handsome yellow toadstool, is most frequently responsible. New rings may be only a foot (30cm) in diameter but old ones can be up to fifty yards (45m) across. Their formation is simple to understand: a single fungus springs up from a spore in grassland and the mycelium grows steadily onwards in all directions, producing over successive generations an ever-widening ring of new toadstools, those on the inside dying back. The largest rings may be several centuries old.

The best edible species are the field mushroom, the cep, chantarelle, shaggy inkcap, giant puffball, parasol mushroom, field blewit, wood blewit, morel, oyster mushroom, St George's mushroom, and the truffle. However, a final word of warning – never eat any fungus unless you are absolutely sure of its identity. Several species are extremely poisonous – the infamous death cap is an innocent-looking white toadstool and fairly common in some woodlands, but it is quite deadly. One cap is enough, so the rule must be: 'if in doubt, don't!'

CHAPTER 2
CHALK DOWN LANDS AND LIMESTONE HILLS

THE lovely chalk downs of southern Britain, with their open rolling hills and short springy turf, epitomise the English scenery just as heather and the Highlands are Scotland. Yet this downland scenery is quite artificial, created initially by man as pasture, and maintained as such for the last 5,000 years by the grazing animals with which he has stocked it.

Towards the end of the last glaciation much of the country was covered by a semi-arctic grassland. As the ice retreated northwards from about 10000BC, trees invaded the country from the south and grassland became quite rare although small areas, grazed by wild cattle and deer, may have survived. The climate warmed a little, and trees were found much higher up the mountains than they are now.

Pollen records show that at about the time of Neolithic man grasses and grassland herbs suddenly reappeared. The main reason for this transformation of the landscape was because the forests were cleared, starting from around 4000BC; the grass and herb seeds would have come from surviving fragments of late-glacial grassland. For instance, we know that plants like rockrose or salad burnet – which have a notably southern distribution today – once grew in semi-arctic grasslands and therefore must have been brought down from their original habitat by the movement of glaciers.

As the forests were cleared and the landscape opened up, the downlands we now know started to become established; there is every reason to believe that large areas had been cleared by 2500BC. It is interesting to reflect, too, that sheep are not native to Britain: they were introduced by Neolithic farmers to graze the new pastures. By the time of Domesday in 1086, nearly all our open downlands were firmly established as pastureland, and maintained as such by millions of sheep.

British downlands divide conveniently into five main areas: the North Downs of Kent and Surrey, the broader South Downs of Sussex and Hampshire, the downlands of Wiltshire and Dorset, the Chilterns of Buckingham and Berkshire and the Wolds of Lincoln and Yorkshire.

Most of these were already open downs in Roman times, and the tops were used as open trackways, providing safer passage than through the wildwood. The Ridgeway path along the tops of the Chilterns and on to the Wiltshire Downs at Stonehenge is one of the oldest highways in the country.

Chalk is a particularly soft kind of limestone which was laid down in shallow seas between 100 and 65 million years ago, formed from huge accumulations of empty shells of sea crustacea. True limestones are composed of both older and harder rocks. Chalk occurs mainly in southern and eastern England, while the harder limestones are deposited in wide arcs away from it to the north and west. Next to the chalk is oolitic limestone, a rather soft, yellowish stone, much in evidence in Cotswold buildings and still used extensively. Farther north and west, in a larger arc, come the hard carboniferous limestones of the Mendips, Gower, Great Orme Head, the Derby Dales and the Craven Uplands of Yorkshire. Elsewhere the limestone areas are small; most of the rest of Britain's

PREVIOUS PAGE
Sheep, the traditional grazing animals of chalk downland, on Old Winchester Hill, Hampshire. As man cleared the forests to provide good pasture, it was essential that it was continuously grazed. Rabbits also contribute to the maintenance of good grazing as they chew off the tips of leaves and so promote further dense growth.

LEFT
Wild mignonette, a typical midsummer chalk downland flower. It flowers from June to September and is found on disturbed waste and cultivated land, on road verges, beside paths and at the edge of fields. It can be self-pollinated or pollinated by bees.

rocks are formed of ancient siliceous material.

Soft contours are characteristic of soft chalk, while harder limestones – even relatively soft oolitic limestones – have frequent cliffs and 'edges'. Some of these are most spectacular, like Gordale Scar and Malham Cove in the Yorkshire Dales, and Cheddar Gorge in Somerset. Chalk cliffs on the other hand, only occur naturally where the sea cuts out the soft rock; the 'white cliffs of Dover', Beachy Head, the Needles, Durlston and Flamborough Head are classic examples of this, and mark the extremities of the chalk in Britain. The 'river cliff' created where the River Mole cuts through the North Downs in Surrey has already acquired softer downland contours.

By 1700 grass and clover seed were recognised items of commerce, enabling the peasant farmer to sow grass as an arable crop, although it was not until the nineteenth century that this became common practice. Throughout the nineteenth century mile after mile of chalk grassland was ploughed up; during and since World War II and with the help of modern technology, we have continued this destruction on a much greater scale. Dorset now has less than one-twelfth of the downland it possessed in 1800; the estimated total for the whole country in 1966 was only 108,000 acres (44,000ha). Furthermore, estimates for 1988 show this may have fallen to little more than 50,000 acres (20,000ha).

Although chalk downland is poor in nutrients the flora is rich, based on the lime content, and it tolerates this deficiency. However, once downland has been ploughed, the original flora never fully recovers. Modern chemicals do further damage – fertilizers, for example, promote the growth of the strongest grasses, which effectively blot out all the other plant species. And herbicides, too, either kill off everything prior to reseeding, or kill the broad-leaved herbs so as to leave just the grasses.

The main influence on the ecology of chalk downland has always been man and his management of the land. For years he has cleared the forests and sought to establish good pasture. Continuous grazing is vital in maintaining these grasslands – grazing by sheep produces the short dense sward required as the grasses adapt to cope with grazing pressure by growing continually from the base; thus when the tips of the leaves are eaten off,

their growth is not really impaired, unlike other plants.

The rabbit was imported by the Normans, but did not become an important animal in the ecology of grassland until well into the eighteenth century; its contribution to the maintenance of open downland was then considerable, until the great crash in its population caused by myxomatosis in 1954. By this time sheep grazing had declined in favour of arable farming on many chalk downs, although the higher limestone hills are still largely devoted to sheep, even today.

If grazing ceases, open downland is soon invaded by scrub, including hawthorn, dogwood, privet, spindle and wayfaring tree. This is clearly an intermediate stage between grassland and woodland, and if no action is taken to clear the scrub, saplings of larger trees soon appear. Eventually these grow tall enough to be dominant, and soon the complete plant succession from open grassland to woodland will have taken place. The type of climax forest on chalk and limestone varies with the geology, climate and topography of the locality, but over large areas of the south it is likely to be beech or oak, while on the northern limestones it may be lime, ash or oak. Some localised areas appear to have climax woodland of yew or box, although it is not clear why this should be.

From a vantage point on one of the southern chalk downs it is possible to deduce the geology of the area by looking at the surrounding landscape and its vegetation. Between the North and South Downs lies a deep trough, the Weald, uplifted here and there in sandstone ridges. Plants and trees in places like Leith Hill in Surrey include much heather, bilberry, gorse and Scots pine, indicating that the soils are lime-free. Ascending the slopes of the downs the vegetation changes showing clearly that these are chalk, rich in lime with characteristic shrubs such as traveller's joy, dogwood and wayfaring trees. In places the white rock is exposed in cuttings, just as the sands of the Weald are often visible, too.

To some extent the chalk vegetation itself is zoned. The lowest areas – which eventually merge with the sandstone – are a grey-white soft rock covered with grey loamy soil. This is known as 'lower chalk', and is easily cultivated. Above this are steeper open grasslands, mostly on white 'middle

Old man's beard – other names include traveller's joy or wild clematis – festoons the hedgerows and trees on chalk or lime soils. It has twining stalks which promote its vigorous growth. In summer the fragrant creamy flowers ripen to clusters of fruit, and in winter the plant develops long hairy plumes.

RIGHT
Malham Cove in the Yorkshire Dales National Park is a classic limestone landscape. The limestone pavement on the plateau above the cove is a world-famous geological formation which attracts thousands of visitors every year, as does the cove itself where a large waterfall, now reduced to little more than a stream, which emerges at the base, once cut back the 350ft (100m) cliffs.

chalk'; while the brow of the scarp may well be the topmost layer of 'upper chalk'.

Most of the flat plateaux on top of the downs are covered by mature woodland, often oak mixed with beech. Where the oak is tall the ground has a good soil layer, lying on a stratum known as 'clay-with-flints' where the chalk has long been leached out leaving a deep soil filled with flinty nodules. This soil is often deep in leaf litter and humus, in some places sufficient to counter the influence of the lime in the chalk beneath. Calcifuge plants may then reappear, with heathers and extensive areas of bracken on the leached acid soil. The oak is a good indicator here; it is stunted and low growing on the bands of chalk because of the shallow soils, and much taller on the deeper soils of the clay-with-flints.

Examine a chalk cutting by the roadside; it will probably demonstrate that the soil, filled with a mass of the roots of closely growing plants, is only 4–5in (10–13cm) deep before the white bedrock is reached. It is this dense mass of plant roots which accounts for the springy feeling of the turf, and which makes it so good to walk upon. Water drains through this mat of roots very quickly and is then absorbed by the chalk to a considerable degree, a cubic foot (0.2cu m) holding a gallon (4.5 litres) of water without appearing to be wet. Excess water filters through the chalk to emerge where it meets the next impervious geological layer, in most instances the gault clay at the bottom of the slopes. This junction of strata holds the source of many of the springs and streams in southern England. Plants on chalk slopes thus have a real problem of water conservation, and meet it by being either low growing like thyme and rockrose, thickly cuticled or downy, or by possessing long roots, like salad burnet which has roots 2ft (60cm) long to reach the water held in the rock itself.

Chalk downland plants are remarkably varied and are nearly all perennials, because competition for seedling space is fierce, and nutrients are too limited for quick annual growth. They in turn pro-

vide for a large and varied population of insects and other invertebrates, of which the many butterflies are the most notable, one of the major attractions of chalk grassland. The abundance of calcium in the chalk also means that invertebrates with shells – like snails and woodlice – are very numerous.

There are no particular birds specific to downland, although grassland species like skylarks, meadow pipits and partridges may be quite numerous. Green woodpeckers hunt the anthills using their long sticky tongues to collect grubs and adult ants. Kestrels and, increasingly, sparrowhawks may hunt the open hillsides from nearby woods – the kestrel to pounce on a beetle or grasshopper; the hawk, in dashing flight, to pluck an unsuspecting pipit from the air. Many species appear on downland in winter – parties of rooks, jackdaws and magpies hunting for invertebrates, while finches search for seeds. Rarer species

such as stone curlew, quail and wheatear struggle for survival in the modern farm landscape.

Chalk is soft and therefore easy to dig, which is why badgers are quite common on chalk downland, their setts usually on a steep slope beneath a canopy of trees or under chalk scrub; they are surprisingly frequent beneath yews. Great spoil heaps of chalk and flints accumulate over the years and an active sett is pretty obvious. Paths radiate down on to the open downland where the badger will find a variety of invertebrate food – worms, snails and slugs – in great abundance.

Limestone is a harder rock than chalk, laid down some 300 million years ago, and formed from the shells of countless marine animals. Despite its hardness and physical strength it weathers easily, the calcium carbonate from which it is made dissolving easily in slightly acid rainwater; tiny fractures in the rock become enlarged by the action of this acidified water into huge fissures, which end up by absorbing all surface drainage. The wonderful limestone scenery and landscape from the Cheddar Gorge in the south-west, by way of the Mendips, the Gower, the White Peak and the Yorkshire Dales to the cliffs of Durness in the far north can be attributed to this simple geological fact. This magnificent scenery is perhaps at its finest in the Yorkshire Dales National Park, centred on Malham.

Limestone hills are therefore often quite without water, as rainwater drains through the cracks and fissures to the lowest levels of the limestone strata to emerge elsewhere as springs – although even these may suddenly vanish, too. The landscape is rugged, dominated by crags, gorges and dry valleys, while below ground huge cave systems may occur. This landscape scenery is called 'karst', named after a region of dramatic geological appearance in Yugoslavia; where the landscape formation has been further affected by Ice Age glaciers, it is called 'glacio-karst' scenery.

The several Ice Ages, which had such a dramatic effect on Britain's landscape, were interspersed with warmer periods known as inter-glacials. During these periods the erosion continued – as it does today – due to the action of slightly acidified rainwater, in the form of dilute carbonic acid, coupled with winter frosts. As the ice returned in successive Ice Ages the underground erosion, which was hollowing out the caves, was frozen into inactivity.

Ice then became the main erosive force, as glaciers deepened the valleys, cut back the limestone scars and scoured the plateaux clean. This clearing of the flat plateau areas by heavy glaciers produced the strange limestone pavements we see today, smooth sheets of rock with their own intricate micro-geology. The individual boulders found on these plateaux are evidence of this glacial activity; they are of different rock types, and have obviously been brought from afar by the ice, and dropped along the pavements when the ice retreated. They are called glacial erratics and may sometimes be found perched on a pedestal of limestone where the surface all around has eroded away.

Limestone pavements are most widespread in the Yorkshire Craven Uplands, with the best areas under maximum legal protection to prevent the removal of any of the stone rockeries. Gait Barrows, Ingleborough, Malham Cove, Whernside and Shap Fell are good examples. Since the retreat of the ice, the solution of the limestone along the cracks and joints has resulted in remarkable patterns; these fissures are called 'grykes' and the paving stones between are called 'clints'. Most grykes are less than a yard deep, although some are as much as a yard wide and five yards (4.5m) deep. Gutter-like features called 'runnels' occur on many clint surfaces; also, hollows in the surface are formed in the same way and are known as 'solution cups'. The flora of limestone pavements is most exciting because many rare plants grow in the stable humid atmosphere of the grykes, protected by the rocks from the grazing sheep.

As the ice cover melted each summer, meltwater rivers gouged out dramatic gorges such as Gordale Scar; and features such as Malham Cove, where a large waterfall once cut back the cliffs from the line of the Mid-Craven Fault, were formed. Now it is no more than a stream, and only a trickle in summer, emerging at the base of these impressive 350ft (100m) cliffs. These dramatic landscapes were formed by the largest and most powerful streams; other valleys, long since dry, were also shaped originally by the action of ice and by lesser streams which have long vanished.

Streams which flow from a higher slope of grit or shale may suddenly vanish when they meet the limestone. These points are known as 'stream sinks' or 'swallow holes' and sometimes provide ac-

A limestone pavement with bloody cranesbill in the grykes, fissures formed in the limestone since the retreat of the ice. The areas between the grykes – the clints – are remarkably flat, and the whole area of these limestone pavements supports a fascinating flora, safe in the stable humid atmosphere of the grykes.

cess to huge underground cave systems. Many of the fells are capped by a layer of hard, impervious millstone grit, so the rainwater gathers into streams until it reaches the limestone; in other places the glaciers have left a deposit of boulder clay on the hillsides, often ten feet (about 3 metres) thick – in places the clay has subsided into underlying fissures in the limestone, producing characteristic depressions known as 'shake-holes'.

Wherever limestone hills occur there are sure to be underground cave systems. Those beneath the limestones of Cheddar and the Gower are famous, and caves can be found beneath nearly every hill in the Yorkshire Dales. They often form an amazingly complex system of tunnels, vertical shafts and chambers, and are sometimes quite vast, extending many miles underground between 'sink' and 'rising' where the water finally emerges. The complexity of any cave system is due to the action of water as it erodes a way out for itself through successively lower routes, until it reappears at the base of the limestone. The dry chambers often have beautiful formations of stalactites, which 'hang' like icicles, from the cave roof, and stalagmites which rise from the cave floor: these are accumulations of carbonate of lime, formed as successive water droplets deposit their dissolved calcium carbonate at the tip; they are coloured by dissolved minerals from the surrounding rocks.

A walk through the hills and vales of this magnificent limestone scenery will reveal all the features which make limestone geology so fascinating. However, the detrimental effects of farming and quarrying, and the erosion caused by the hordes of visitors, constitute a growing problem: the natural beauty of all of Britain's fragile limestone areas is now at risk.

OPPOSITE
Bloody cranesbill and English stonecrop growing together in a limestone crevice. The bloody cranesbill, an indicator species of lime-rich soil, produces its brilliant purplish-red flowers from June to August. English stonecrop is a deep-red creeping perennial which forms mats of dense leafy stems. Its white flowers are produced from June to September.

Wildflowers of Chalk Downs and Limestone Hills

The slopes of chalk and limestone downs are full of colourful flowers in early summer; there may be as many as forty different plant species in a square yard of turf.

Chalk and limestone hills have many species of flower in common because both consist of calcium carbonate. The shallow soils are rich in humus and calcium, but lack nutrients such as phosphorus, nitrogen, potassium and iron. These conditions favour slow-growing species which form a dense sward, rather than the taller, more rapid and vigorous growers – in any case these are grazed to sward level by sheep. Most chalk plants are perennial, since annuals find it difficult to gain sufficient foothold or quick nourishment in the dense plant covering.

There may be only a couple of inches (5cm) of soil cover before bedrock is reached, and most plants have their roots in just these top two inches (5cm). Only woody plants, like common thyme and St John's wort, extend long roots deep into the chalk to find water; these are then able to withstand hotter conditions and flower later in the summer. Chalk and limestone soils are warm, dry and very well drained. Most plants have adapted in various ways, to withstand these dry conditions and in order to conserve water; for example, many flower in May and June and are dormant in the hottest months. Many are low growers, keeping most of the plant in the inch or two (2–5cm) just above the ground, and so living in the humid microclimate which prevails at soil level. Most have small, thick leaves, often in a basal rosette, and these are often downy. The rosette habit is adopted by plants like hawkbits, stemless thistle and hoary plantain, and gives protection against grazing animals. More succulent plants such as orchids flower early, and have withered by the hottest part of the summer.

It is relatively easy to discover whether the base

soil is lime-rich. Certain flowers avoid lime altogether, foxglove and broom being classic examples; these are called 'calcifuges' (from the Latin *calx* – chalk, and *fugio* – I flee). 'Calcicoles' are plants which must have lime soils (from the Latin *calx*, and *colo* – I inhabit). Among the indicator species, the most conspicuous are common rock rose, clustered bellflower, small scabious, kidney vetch, bloody cranesbill, dropwort, hoary plantain and yellow-wort. Southern calcicoles include traveller's joy, dark mullein, dogwood and wayfaring tree.

Orchids are perhaps the best-known flowers of chalk and limestone, and over half of the fifty or so species of British orchids are characteristic of lime-rich soils. Orchids are all rare by comparison with most other wild flowers, and many have full legal protection. The seeds, produced in thousands, are microscopic and grow very slowly in association with a fungus which lives on the roots, an association called a mycorrhiza which is essential to the orchids' well-being. A bee orchid takes eight years to grow from seed to flower, while a twayblade may take twenty years. Some, like the spotted orchid, may flower for several years while others, like the bee orchid, flower once, set seed and die.

First to appear in April is the early purple orchid, found in woods and on the open downs; this is followed in early May by green-veined, fly, man and burnt-tip orchids. June is the most productive month when many are in flower including spotted, fragrant and pyramidal orchids, three species with tall flower spikes of pink or red flowers. Much rarer species are the lady orchid, found in only a few Kent nature reserves, and the soldier and monkey orchids, found in only one or two localities.

The bee, fly and spider orchids have just a few

LEFT
A bee orchid awaiting pollination. The striking similarity of the lower lip to a female bee helps to attract the male. It flowers from June to July. Other orchids – the fly, early and late spider orchids for example – also resemble the females of the species which pollinate them.

RIGHT
Cowslips on the short chalk turf of the South Downs, Sussex. These deep yellow flowers with orange markings in the centre are found in grassy habitats, meadows and pastures, in scrub and open woodland, banks and roadsides. The cowslip is the foodplant of the Duke of Burgundy fritillary and flowers from April to May.

striking flowers, rather than a dense spike of small florets, and are largely of southern distribution; there are many more related species in southern Europe. The lower lip of each flower resembles the female of a particular species of insect, while some species even produce a scent which resembles that from the female insect, thus attracting the males which, in visiting one flower after another, effectively pollinate them.

There is wide variation in climate between the warm, dry, southern chalk and the colder, wetter, northern limestones. Flowers like bird's-foot trefoil, horseshoe vetch and clustered bellflower grow just as well on the northern limestones as they do in the south, but others are far more particular. For example, the wetter conditions of the Craven uplands of Yorkshire favour globeflower and bird's-eye primrose, but these would never grow on the dry South Downs. Similarly, maidenhair fern grows on the limestone of the West Country and Ireland, but would not tolerate the frost of the south east. Clearly climatic conditions are vitally important, but this is not the sole reason for such striking differences between the northern limestone flora and that of the southern chalk. Much of the chalk is very pure, and may reach 95–8 per cent pure calcium carbonate; limestones are less pure, laid down in a mixture of sand and mud millions of years ago, and are also harder. It is because of these differences that limestone possesses certain characteristic species which are not found on the softer chalk.

Perhaps the rarest of the northern flowers is the lady's-slipper orchid; it may now even be extinct, but would be found in late spring in light ash woods. Another rarity, and an orchid characteristic of limestone rocks and pavement, is the dark red helleborine which flowers in late June; plants such as grass of Parnassus, bird's-eye primrose, bitter milkwort, Jacob's ladder and angular Solomon's seal also prefer the limestone pavement. Many other plants such as the mountain pansy, lady's mantle, mountain cranesbill and northern bedstraw are found quite frequently on the northern limestone but never on the chalk.

Likewise there are many plants which are more characteristic of the southern chalk, where the downlands change their colours according to the season: from blue in late spring, when hoary and chalk violets and milkworts flower in profusion; to yellow in midsummer, with lady's bedstraw, bird's-foot trefoil, kidney and horseshoe vetch, rock rose and cowslip; to the purples and reds of late summer when the downs are filled with the scent of thyme, marjoram and basil, and butterflies flock to the purple flowers of knapweed and scabious.

There are two creatures very common to chalk downland whose habits and life-style give it a distinctive appearance and substantially affect the surrounding flowers: the rabbit and the the ant. Rabbits maintain an extremely short turf by constant grazing, and certain plants which are habitually associated with rabbits are often found here; these must be plants which benefit from the manuring of the soil from rabbit droppings. Elder is one such plant, and when found on a chalk slope it usually denotes an old warren; henbane, deadly nightshade, stinging nettles and hound's-tongue are all firmly avoided by the rabbit, being either poisonous or at least irritant – thus they survive ungrazed round downland rabbit warrens.

The little anthill mounds which are one of the most prominent features of chalk turf are the work of the yellow field ant. Each dome conceals a honeycomb of passages and chambers, which is home to some twenty-five thousand ants. A broad sweep of chalk slope may accommodate hundreds of such anthills, and moreover their frequency is a good indicator as to the antiquity of the grassland. Some of these anthills may be two hundred years old! They often support a specialised flora, since they offer conditions which are ideal for creeping plants like common thyme and eyebright.

In the south, the most famous flower of the chalk is perhaps the pasque flower, now very rare and found only in a few localities, but particularly favouring ancient earthworks. Like most downland flowers, it disappears completely when grassland is 'improved' for agriculture.

Skylarks are one of the most numerous grassland birds, found throughout the British Isles; the beautiful liquid song is one of the true sounds of summer, uttered in flight as it soars and hovers. The skylark nests on the ground, in sand dunes and marshes as well as moors and fields.

Birds of Downs and Grasslands

In spring and early summer you will *hear* the bird which is most frequent in these rolling uplands before you see it: the ubiquitous skylark. It is found throughout the British Isles, sings only in flight, and nests on the ground, deep in a grass tussock. Its numbers are thin and fairly scattered across the open downs, but in areas of good habitat there may be one pair per 2½ acres (1ha). The total British population of skylarks is thought to be in excess of 2 million pairs. There are two broods and sometimes three, with clutches of three to five eggs most usual. In common with many ground-nesting birds, the chicks leave the nest very early, from nine days old, and hide in the dense grass. They take about three weeks to become fully fledged.

The stone curlew was once found throughout the chalk and heath areas of Britain as far north as Yorkshire; now, however, it is confined to dry heathlands in East Anglia and the wilder chalk uplands of central southern England. Its population was estimated at 2,000 pairs in the 1930s, but by the mid-1980s this had fallen to about 140 pairs because of the pressure of more efficient farming methods. It is a summer visitor, arriving in March and departing in September or October, and prefers to nest in bare stony patches of high downs, among cereal crops and on old heathland. There are usually two eggs in a clutch, incubation takes twenty-five days, and the chicks are 'nidifugous' – that is they leave the nest on hatching. They take six weeks to fledge. The stone curlew is a shy, secretive bird and you are most likely to see it in flight, especially in the evening, when its characteristic curlew-like call carries far across the fields. It is related to the waders, but has adapted to dry conditions; it is more numerous in southern Europe.

The thicker cover around the edge of any downland field may well conceal a covey of partridges, rotund little birds which whirr away on fast-beating, stubby wings. Both the red-legged partridge and the common, or grey partridge are found in this habitat; the red-legged has been widely introduced since 1790, and is still put down for sport by shooting estates. The common partridge has declined throughout Britain and is now scarce in some places. More efficient farming methods and particularly the use of pesticides which deplete the insect supply for the growing chicks are mainly responsible.

The quail, another downland resident, is the smallest European game bird, about the size of a starling and is very often found among cereal crops. However, it is now very rare with only a few pairs ever being recorded each year – although it may be that it is easily overlooked due to its crepuscular habits and small size.

Sometimes a pair of wheatears may be seen popping in and out of an old rabbit hole on one of the quieter downland hills; these lovely little thrushes once nested on many of the downs of southern England, but have now largely disappeared from the area, although they are still common on the higher uplands of the north and west. Their dramatic decline in the south can be attributed to the destruction of the short downland turf in favour of arable fields, coupled with the increase in scrub after myxomatosis decimated the rabbit, an important grazer, in the 1950s. The female lays five to six eggs, incubation is fourteen days, and the birds are 'nidicolous' – that is, they stay in the nest until they can fly.

Where open grassland is interspersed with areas of scrub, the bird population increases considerably. This is a favourite habitat of several species of

warbler, all summer visitors arriving in April from Africa. The whitethroat is the most characteristic scrub warbler, nesting a foot or two (30–60cm) above the ground where grass, nettle, bramble and scrub all meet. Its numbers were decimated in the 1969 Sahel drought, when 80 per cent of those which would normally have returned had apparently died in Africa. Nearly twenty years later the population in Britain is still only half what it was in the 1960s, although it is slowly increasing, its rattling song and fluttering display flight more frequent each year.

Other warblers found in this scrub habitat include willow warblers, lesser whitethroats, and occasionally chiff-chaffs and blackcaps. Tree pipits, also summer visitors, will nest in long downland grass if they have a high tree nearby to use as a song post. Meadow pipits, birds of the higher grass moorlands, are less common on the southern downs, but are still plentiful provided the slopes are of rough grass, and that the level of disturbance is not too great.

Many birds pass along the downland ridges during their spring and autumn migrations as the short turf provides an excellent supply of insects. Wheatears, whinchats, pipits and finches all occur in some numbers. However, as autumn changes to winter the downs lose most of their passerine species, either back to Africa or, in the case of the skylark and the meadow pipit, to the nearest estuarine marshland where winter food is easier to obtain. Flocks of greenfinches, linnets, chaffinches and sparrows may be seen searching in the arable downland fields for spilled grain. Winter thrushes pass through the scrub in search of berries, where redwings and fieldfares from Scandinavia join the local blackbirds and song-thrushes.

A cock blackcap, a summer visitor from Africa, feeds its chicks deep in a bramble thicket. The male's black and the female's brown crown distinguish it from other warblers. Increasingly it winters in Britain and is sometimes seen at bird tables.

Butterflies of Chalk and Limestone Downs

Chalk and limestone downlands provide one of the best habitats for butterflies, and it is not unusual to see some twenty species during the course of a season. Some of the best, and largest, areas may have over thirty – that is, nearly half of all the species of butterfly seen regularly in Britain. On the whole, these islands are not very good for butterflies because the climate is too cool and wet; several of the species present here are found in far greater numbers in continental Europe and are at the extreme north-western edge of their range in Britain. Downlands, however, offer better conditions because they have a wide variety of plant species, especially nectar-bearing flowers, and because they offer a warm, sunny, open habitat.

Most butterflies are very restricted in their choice of food plants for their caterpillars, and many will use only one plant species, though a few may choose between two or three. Normally, downland butterflies lay their eggs directly on the plant. The exception is the marbled white which scatters its eggs among long grasses. The female butterfly will carefully 'test' a plant by 'drumming' on the leaves with its feet; it is assumed that this releases enough chemical 'scent' from the plant for the butterfly to identify it correctly. Butterfly eggs themselves are remarkable, and often have distinctly sculptured shapes, according to the species.

Most prevalent among the chalk butterflies are the 'blues', with several species possible at the best sites. Most blue butterflies have a life-cycle closely interrelated to chalk plants and to other downland invertebrates – most often the ant – and are not found anywhere else. They are from the Lycaenidae family of butterflies, which also includes the hairstreaks, brown argss and the small copper, and the common blue is the most widely distributed. Its food plant is bird's-foot trefoil, a common plant not solely confined to downland which explains the species' wide distribution. The

LEFT
A male common blue butterfly on one of its foodplants, the bird's foot trefoil; it also feeds on clovers, medick and rest-harrow. The most common and widespread of the British blues, they live in colonies. Out of the sun, the butterflies tend to rest head downwards on grass stems close together, sometimes several to a stem. The first brood is seen from May until July, and there is often a second brood in September.

RIGHT
The dark green fritillary butterfly, a strong-flying species of open downland in late June. It is also found in woodland, but is quite at home on moors or exposed cliffs. Wild violets are its foodplant, though in captivity the larvae will feed on the related pansy. The butterflies are also particularly fond of thistle flowers.

has two broods, one appearing in mid-May and flying throughout June, the second from late August until the end of September. Like most blues, only the male has the distinctive powder-blue colour; the female is brown, although some females have a dusting of blue.

Chalkhill blues and Adonis blues lay their eggs on horseshoe vetch and are thus restricted to chalk and limestone areas. Both are southern insects, the Adonis blue especially so, and with the fragmentation of chalk downland by farming, both are now declining. Furthermore, individuals rarely stray far, often spending all their lives in one part of a meadow, thus inhibiting the colonisation of any other areas. Populations are becoming increasingly isolated, although research by mark-and-recapture techniques shows that levels can vary widely from year to year, dependent upon the weather, and that this need not always be to detrimental effect. Thus on one Dorset down the Adonis blue population was estimated at only fifty adults in 1977, but at over 50,000 in 1982.

Adonis blues have two broods, in June and late August, but chalkhill blues are late summer butterflies, appearing in late July. The large blue became extinct in the wild in Britain, but the Nature Conservancy Council is trying to reintroduce it; formerly it was found only in the milder parts of the South West, and was on the extreme edge of its European range in Britain. The large blue depends almost entirely upon symbiosis with the ant for its survival, as the ants carry the caterpillar right into their nest where, in return for honeydew, it feeds on live ant grubs. The large blue is quite common on the Continent, especially in southern France, Spain and Greece.

The small blue is the other regular chalk blue butterfly, appearing in mid-May and feeding on the flower-heads of kidney-vetch, but only to be seen for a short period, flying for no more than three weeks or so. Others, like the small copper, produce several broods throughout the season and may be seen from late April till early October. Despite its brown and scarlet colours the small copper is closely related to the blues, as its under-wing markings indicate. It is not confined to downs since its larvae feed on docks and sorrels. The brown argus, however, is a brown 'blue', feeding on rock rose and storksbill and therefore restricted to the downs, dunes and limestone areas where these plants occur.

The caterpillar of most blue butterflies has a honey gland on its tenth body segment, and the purpose of this gland is to attract the ant. The ants use the sweet drops of honeydew for their grubs; in their turn, they may move the butterfly larvae to plants near their nest and so help keep harmful insects at bay. This mutually beneficial relationship is called symbiosis.

Several other families of butterflies, including the skippers, browns, whites and fritillaries, are well represented on the chalk. Brimstones are often the first to appear, in March, their larvae feeding on buckthorn scrub on the woodland edge between the woods and the open downs. Other early species include the dingy and grizzled skippers, from late April, which lay their eggs on bird's-foot trefoil and wild strawberry respectively. Green hairstreaks are scrub butterflies which appear from early May, their caterpillars feeding on gorse, trefoil and rock rose.

Small heath, wall, small, common and Adonis blue, brown argus and Duke of Burgundy fritillary – all these appear in May; while mid-June should see many grass-loving butterflies including meadow browns, ringlets, gatekeeper and small and large skippers. Marbled whites and dark green fritillaries are a feature of late June, the marbled white especially being very numerous in certain favoured localities. Despite its name, the marbled white is a member of the 'brown' family, the Satyridae.

One of the last species to appear is the comparatively rare silver-spotted skipper, which feeds on sheep's fescue grass. It has declined in its southern downland haunts because so much grassland has been ploughed, and because open downs have been invaded by encroaching scrub, largely the result of insufficient grazing by rabbits after myxomatosis had decimated the rabbit population.

Statistics show that butterflies of nearly all species are declining rapidly in the modern world, primarily because more of their habitat is destroyed every time ancient grasslands, downs, woods, marshes or moors are ploughed up and converted to farmland. It is vital to remember that while one thoughtless collector can do lasting damage to a small colony of butterflies just hanging on to its habitat, several collectors could exterminate the colony altogether.

Other Invertebrates of Downs and Grasslands

In summer the flower-rich downs and grasslands buzz with activity. Butterflies are the most obvious and can be seen everywhere, but there is a huge variety of other insects and invertebrates present.

Ancient grasslands which have not been sprayed provide a particularly good habitat for moths. The brightly coloured six-spot burnet is the most numerous of the burnet moths, found in loose colonies and usually on chalk downland. The larvae feed on vetches, clovers and especially bird's-foot trefoil. The moth flies freely from flower to

The common field grasshopper can often be seen in rank vegetation in midsummer. With the meadow grasshopper, it is our commonest species. Both are colourful shades of green and brown and are found throughout mainland Britain.

flower and is rarely molested by birds because it emits several unpleasant substances if attacked, including hydrocyanic acid – cyanide! This it obtains from clovers and trefoils while still a caterpillar. Its larvae spin papery pupal cases high up on grass stems, conspicuous but out of reach of birds as they cannot settle to remove them.

The small, metallic-green forester moths are also daytime chalk-down moths, and closely related to the burnets; there are three species. Cinnabar moths are notable for their coloration – they are red and black, and their larvae, which feed on ragwort, have orange and black stripes; both are examples of strong warning coloration. Other common day-flying moths of downs and scrub edges include the wood tiger, latticed heath,

speckled yellow, white plume and green longhorn moths.

Grasshoppers and bush crickets are also notable members of the invertebrate fauna, much in evidence in downland grasses from June till October. There are eleven species of grasshopper, and all except the meadow grasshopper have well-developed wings and fly well. They can be seen eating grasses and other low-growing plants from about midsummer onwards. The females lay batches of eggs in the soil and these overwinter, hatching in spring. Young grasshoppers and bush crickets are called nymphs; they look very like the adult but are much smaller, and their wings are not developed. They experience four skin moults whilst growing to adult size. All of them, nymphs and adults, hop-

pers and crickets, can jump very well with powerful back legs.

Although both grasshoppers and bush crickets are of the order Orthoptera, they have notable differences. Grasshoppers have short, thickened antennae whereas those of bush crickets are long, thin and flexible. Female bush crickets have a curved ovipositor (egg-laying organ) which is not present in true grasshoppers. The hearing organs of true grasshoppers are situated on either side of the abdomen near the thorax; those of bush crickets are in the front legs. Bush crickets live on scrub vegetation and eat both plants and other insects. On the back legs of the grasshopper there are tiny, hardened projections which are used to produce the familiar 'buzz' or stridulation heard in midsummer grasslands – this noise is made by rubbing the hardened veins of the forewings over these knobs on the back thighs, and is used by the males to attract females. Bush crickets stridulate in a different fashion, by rubbing both forewings together. Common species of grasshopper include the meadow, field, green, mottled and stripe-winged; and of bush cricket, the great green, dark and speckled.

There are over six hundred species of spider in Britain, many of which can often be found in grasslands and on downs, all equipped with a bite which, while quite harmless to humans and carnivores, is poisonous to their tiny prey. Most spiders can spin a silken web, produced from spinarets on the abdomen; the silk is extruded as a liquid from the silk glands but solidifies rapidly in air. There are many different types of web, but all are designed to catch prey – circular orb-webs are the most familiar, placed in insect flyways and enormously strong, capable of holding quite large insects. Some spiders build sheet-webs which are constructed flat in the vegetation or on the ground to catch crawling insects; others sit below a trapdoor of silk. Some hunt their prey by running and jumping and do not make a web at all. Spiders are not able to digest a

meal internally, so they inject digestive juices into their prey and then suck up the food in liquid form. They are often beautifully coloured, especially those that live in flower-heads. These are called crab spiders because of their shape, and their colours have evolved to blend with the host flower so that they can lurk undetected, waiting for visiting insects.

Where downland has a good variety and covering of flowers, bees and wasps may be seen in reasonable numbers. They both belong to the insect order Hymenoptera (literally 'membrane-winged'), along with sawflies, ichneumons and ants – in all no fewer than 6,200 British species. You will see many honey bees, but they comprise only one of these species. There are some sixteen species of bumblebee, which differ from honey bees in that their colony is annual – only the young fertilised queen survives the winter. Besides these there are cuckoo bees, which resemble bumblebees and invade their nests; leaf-cutting bees, which cut chunks out of rose leaves and use the segments to line the cells of their nest; and also solitary bees and wasps, which live quite alone in small tunnels or cells. A hive may carry 50,000 individual honey-bee workers, while a bumblebee nest is much smaller, with only some 200 individuals. Common wasps may reach 3,000 in one nest.

Ants are also part of the order Hymenoptera, and there may be millions colonising a typical downland slope. One colony alone of the yellow field ant may consist of as many as 20,000 individuals, although this is a very common species on downlands, responsible for the numerous, vegetation-covered anthills. Ants are predatory insects and capture a wide variety of food items from among their fellow invertebrates.

Calcareous grassland, with its abundance of lime, is particularly good for snails as it supplies the calcium they need to build shells; old quarries are especially good sites. However, all snails need moisture, and this is often a problem on chalk; several species aestivate in hot summers, attaching themselves to the tops of tall stems away from the hot ground and sealing themselves up. Common species on chalk downs include the huge Roman snail (an introduced species), garden snail, round-mouthed snail, ribbed snail, door snail, wrinkled snail and banded snail.

Crab spiders usually hide in flowers to pounce on their insect prey. One of the most numerous crab spiders, *Misumena*, is able to alter its colour from white to yellow or even pink in order to resemble its host flower, while the male of the species, *Micrommata*, has bright yellow and scarlet stripes.

CHAPTER 3

HEATHLANDS

Large tracts of sandy heathland once covered wide areas of southern England; indeed they probably covered most of the area between the ranges of chalk hills in south-east England, and then northwards into East Anglia and Yorkshire. Perhaps surprisingly these heaths were all originally created because of man's influence, the result of wide forest clearances from about 4000BC onwards. Pollen studies at Hockham Mere in Breckland, Norfolk, and elsewhere, show that before 6000BC, areas which we now consider to be traditional heaths were in fact covered by woodland. Yet there seems every reason to believe that man was managing all these areas as open space even before Roman times, and all were in place long before the time of Domesday in 1086.

Man was responsible, because clearing the forests ultimately caused a change in soil structure, from the rich organic woodland humus to the thin sand of the heaths we know today; furthermore, his practice of constant grazing with stock animals maintained this heathland habitat, as over the centuries it effectively suppressed the tree seedlings once and for all. Burning, too, was often a policy specifically employed in pasture management, especially in spring when the winter heather and gorse was dry, to promote new and vigorous growth for cattle and sheep to browse.

After the arrival of the Normans, heathland was used extensively for farming rabbits, and warrens were created all over England and into the Scottish Lowlands; the largest number was to be found in Breckland, East Anglia. The animals were confined within a certain area as well as possible by a great bank and ditch, often topped with gorse, and the boundaries patrolled. The earliest warrens were on islands – the Scillies, Skokholm, the Farnes – which afforded further protection to their valuable inmates, and greater ease of management. Rabbits were prized for their fur and meat in the Middle Ages, and rabbit farming was highly organised.

Heathland soils are sandy, acid and poor in nutrients, the clearance and suppression of the trees so many centuries ago having changed the structure and chemistry of the soil in these areas. Denied the beneficial effect of the trees' extensive root systems and the nutrients supplied by a woodland's organic cycle, the soil became thin and impoverished. Without the protective layer of leaf humus and forest cover, the underlying rock typical of these areas was left relatively exposed and was readily weathered, creating over the centuries some of the heath sands we see today. Furthermore, when burning was

PREVIOUS PAGE
The stone curlew, once found on many heaths and downs, has declined to under 200 pairs in twenty years due to loss of habitat. It is a summer visitor, usually nesting on bare open ground or ploughed fields in a scrape lined with small stones or rabbit droppings.

Typical heathland with heather and bracken – a product of prehistoric forest clearance. A wide variety of heaths and heathers is found throughout Britain in areas where they are specially adapted to survive in impoverished soil. The shades of white, pink, mauve and purple provide a stunning sight, interspersed with the sharp bright green of new bracken.

uncontrolled, and too widespread and fierce, it seriously damaged the vegetation, often destroying large areas of heather and gorse completely so that nothing would grow on the scorched ground for years.

Once an open heath has become established, heathers have the capacity to produce a distinctive soil type known as a 'podzol'. The surface is covered by a slowly rotting layer of leaf mould which is high in acid. Rain drains quickly through this and the sandy soils, and because of the acidity, dissolves and takes with it most of the nutrients from the sands, leaving a pale, whitish-grey, impoverished soil. Several inches (some 5–10cm) below the surface there may be a hard, black layer, called a 'humus pan', and well below this there is often a very hard, rust-like layer called an 'iron pan'. The rainwater has therefore effectively leached the organic salts from the humus and the iron-rich minerals from the sand, leaving behind the characteristic soil profile of heathland.

Some heath sands as we have seen, have been produced by the weathering of the underlying rock; others were drifted into place thousands of years ago by wind, rivers and glaciers – in the south and east much sandy heathland lies on a bed of solid chalk. A few inches (5cm) of drifted sand are enough to produce acid soils and heath vegetation, though not a 'podzol'. This mix of geological structures is the reason for the mixture of chalk and heath flowers to be found in the Brecks of East Anglia and places like Newmarket Heath.

True heathland, now largely destroyed in western Europe, is a precious habitat *because of* the impoverished nature of the soil, and because of the rarity of the limited variety of species which thrive on it. Many of the plants and animals living here are highly adapted to survive in difficult conditions, and would not survive elsewhere. The preservation of heathland is now a matter of urgency, as urban developers turn their attention to the wide open, easily cleared spaces around many southern towns and villages. Ironically, the very nature of heathland means that conservation, so often an attempt to improve a habitat impoverished by man, must in this case try to maintain the poverty of these soils, and fight against enrichment. Burning may still be used as a management tool, but it must be a controlled, 'cool' burn which does no more than lightly scour the heath with fire, perhaps carried out in strips to rejuvenate the heather, whilst leaving older thickets for Dartford warblers and song posts for woodlarks.

Heathland Flowers

From early July to late September Britain's lowland heaths are filled with acres of purple heathers. Any of these heaths may well be bounded by woods of Scots pine, and here and there clumps of gorse and groves of birch will be prominent. Otherwise, however, heath flowers need to be searched for – they are there, but the species variety is small compared, for instance, with chalk downland. A survey of flowering plants may well reveal thirty to forty species in a square yard of chalk turf, but only three or four on the heath.

The heaths of Surrey, Dorset, Hampshire, Norfolk and Suffolk are found on light sandy soils which provide nutrient-poor, acid conditions similar to those on many upland moors. The dominant family of flowers on both habitats is the heath family Ericaceae. Heaths and heathers clearly go together, the words themselves having altered little in meaning since Anglo-Saxon times. There is, however, no clear distinction between lowland heaths and upland moors. Both resulted from forest clearance, but the climate of lowland sandy heaths is usually dry and even of periodic drought, while moors are in areas of high rainfall and are usually based on peat and hard, acid rock. Furthermore, as the open spaces of lowland heaths are man made, they would quickly revert to woodland if they were not managed in some way. So would some moorland, if grazing stopped.

Heather is quite widely distributed, from the western Mediterranean as far north as the Arctic, but it is most prominent and luxuriant in the north Atlantic fringe of western Europe. In the harsher climates of central Europe it is restricted to sheltered open woodland.

Several species of heather are found on lowland heaths, but common heather or ling is most dominant here; it seems particularly well adapted to our cool oceanic climate, its many tiny leaves able to collect light from all directions when the sky is overcast or cloudy. The leaves themselves are small and leathery, and this structure helps to conserve the plant's water intake in a dry environment. Heather grows well on wet peat, but it dries out quite readily, both in hot summers and if the peat is frozen in winter; it may also suffer in cold winds. It demonstrates a true colonising ability and will quickly cover large areas of dry sandy soil, as it did after the forests were felled. It produces huge numbers of tiny seeds and strangely enough, it benefits from heath fires which assist the germination process.

On the driest parts of the heath, and especially on sites of recent burning, the most notable plant is bell heather, with a more showy, deeper purple flower than common heather. At the other extreme of this habitat, where dry heath sinks down to wet bog, you will find the third common heather species – cross-leaved heath. This thrives on the wet, acid soils which are so often found on heaths, soils created as such because of the impervious layer of iron stone which has leached and formed below the surface, and which results in bog conditions occurring above it.

There are several species of gorse prevalent on heaths. Common gorse is the most conspicuous of these, growing into bushes that may be 6ft (2m) or more high. It fills the heath with a blaze of deep yellow flowers in late spring, and continues to flower intermittently thereafter. Dwarf gorse is much smaller, flowering in high summer; petty whin is also a small plant, and very sharply spined.

Several heathers and gorses will only grow in the West Country, extreme examples of plants which require a particularly Atlantic-climate environment. Dorset heath is a larger, strikingly beautiful flower found in the Isle of Purbeck, while Cornish heath is very rare even in the South West, growing mainly on the Lizard peninsula; finally, St Dabeoc's heath is found only in south-west Ireland. Western gorse is another which is more widespread in the Atlantic counties, and flowers in autumn.

Here and there among the heather there may be a variety of common plants growing side by side, traditionally associated with each other and with heathland and including tormentil, heath bedstraw, heath speedwell, lousewort and sheep's sorrel. Most of these are found in grassy patches alongside paths and tracks. Other common plants include common cow-wheat, heath milkwort, heath woodrush and wood sage, while the characteristic orchid is the heath spotted. The grasses are usually sheep's fescue, wavy hair grass and fine bent.

In the immediate aftermath of high-summer heath fires, the specialised conditions which prevail, albeit only temporarily, are perfect for both bracken and rosebay willow-herb and large stands of both plants may occur. Another name for rosebay is 'fireweed'.

Also characteristic of dry heathland is dodder, a strange parasitic plant whose leafless, threadlike stems cover the heather and gorse in long red strands which bear little pink flowers later in summer. Because it has no leaves, it is unable to produce food itself but obtains what it needs by way of its entwining tendrils which pierce the host plants.

Some of the most striking heath flowers are found in the wet conditions of heath bogs. Early flowers may include marsh orchids in May and June, while cross-leaved heath is now the dominant heather. The most striking feature of early summer are the fluffy white heads of cotton-grass waving in the breeze, and heathrush, pill sedge and purple moor-grass are also common. Sundews are often numerous, bearing sticky hairs on their leaves which are designed to catch insects, the flowers thereby compensating for the lack of nitrogen in the soil by catching their own. Butterworts may also occur, although they are more common in the north and west; and these, too, obtain the extra nutrients they require by catching insects on their leaves. There are other flowers which occur on the wetter heaths of more northerly and westerly distribution, including cranberry, bog myrtle and bog rosemary; all of these are comparatively rare.

Yellow tormentil and pink lousewort often grow together on dry heathland soils. Tormentil is in flower from May to September and is very common on acid grassland, heaths and moors. Lousewort flowers from April to July.

As spring changes to summer, heath pools may be carpeted in white bogbean which grows actually in the water but raises its white flowers above the surface. Bogbean, flowering in June, gives way to thousands of golden flower spikes of bog asphodel in July, the flowers being followed by the bright red fruiting spikes in August.

Much rarer flowers include the tiny bog orchid often found beneath clumps of cross-leaved heath, and the showy marsh gentian, both flowering in late summer.

The heathland flora may lack variety, but its flowers form a distinctive and colourful addition to the rather stark habitat.

Heathland Birds

Heathland birds comprise a fairly specialist community. Heath is a basically unproductive habitat, especially where heather is dominant, and even in summer the bird population is sparse. However, where the heather is mixed with a variety of shrubs, species found elsewhere will come and breed on the heaths too, and the bird population then increases.

Some species of bird are found only on heathland, the nightjar and the Dartford warbler being the most typical. However, as large tracts of lowland heath have been destroyed for agricultural purposes and for suburban development, so its unique wildlife, and especially its birds, have declined even further – the Dartford warbler was always a comparative rarity, being at the northern edge of its European range in southern England. It used to be our only resident warbler, the rest migrating to Africa in winter, but the marshland Cetti's warbler has begun to breed in very small numbers in the southern counties. Dartford warblers eat insects and therefore just manage to scratch a living in the cool British winters, though hard weather greatly reduces the small population, which is probably fewer than 600 pairs. A few pairs occur on the heaths of west Surrey, but the majority are found in the New Forest and on the Wessex heathlands of Dorset and Devon. The Dartford warbler prefers a mixture of old heather and mature gorse, the male using the tops of the taller gorse bushes as song posts in spring; its nest is usually deep in the heather just above the ground. There are normally two broods of four eggs, incubation taking twelve days and the young taking about fourteen to fledge.

The nightjar is the other species of bird unique to heathland. It is essentially a dusk-time bird and therefore hard to find in daylight, and is more likely to be heard first, rather than seen – as the sun goes down in midsummer its peculiar call will carry distinctly over the heath, a low-pitched, grasshopper-like, constant 'churring' which may last for several minutes on end. In flight it darts, flaps and glides like a large dark moth; the male has noticeable white patches on its long, hawk-like wings and tail.

The nightjar has declined over the past fifty years to the point of rarity. Its numbers are thinly scattered over many counties as far north as the Scottish central Lowlands, but it is most readily found on these lowland heaths where it prefers open heather with scattered clumps of birch and pine. It has recently been found breeding in young, second generation conifer plantations which may help to slow its decline; its total British population is only about 2,000 pairs.

Nightjars arrive from Africa comparatively late, in mid-May, and usually lay two well-camouflaged eggs on bare ground between heather clumps. The adults feed on insects caught in flight, returning at intervals to feed the chicks, also perfectly camouflaged. Incubation takes eighteen days, as does fledging, and the young are independent at about six weeks. They often have two broods, but are still on their way south by mid-August; they will have spent only three months in Britain, but nonetheless will have taken full advantage of our brief, insect-filled summer.

The stone curlew is another bird found on heaths and downland, but in both habitats it has suffered badly from intensive modern farming methods. On heathland it occurs mainly on the Brecks – open stony ground with scattered heather and gorse which suits its needs admirably. Like the nightjar it is a summer visitor from Africa, but arrives as early as March. Its population is a little above 100 pairs.

Both the woodlark and the hobby are also rare birds largely confined to heaths; the woodlark population stands at hardly above 200 pairs, and the hobby at twice this figure. Like other heathland birds the woodlark has declined dramatically in the past fifty years. It is a secretive little bird, found singing from treetops beside heathland tracks, where it likes to feed on the short pathside turf. It is resident throughout the year but, like the Dartford warbler, is on the northern edge of its European range here and hard winters cause it great difficulty. The nest is well hidden close to the ground, and is quite one of the most difficult of all British birds' nests to find. There are normally two broods, with incubation and fledging (like that of many small birds) taking about twelve to fourteen days.

The hobby's nest, unlike the woodlark's, is often easy to locate, but is invariably built in the top of the highest Scots pine available – a most fortunate habit. The hobby, a small falcon, is most often found on the heaths of southern England, although a few pairs breed in isolated downland woods westwards to Wiltshire, Somerset and Devon. It arrives from Africa in mid-May, lays two to four eggs in mid-June – usually in an old crow's nest – and incubates them for twenty-eight days; the chicks hatch in mid-July and fly in mid-August.

The best time to appreciate this superbly aerobatic little falcon is just after its arrival in mid-May when pairs indulge in dramatic display flights over their territory, especially early in the morning. Late July sees another burst of activity as the chicks begin to grow larger and both parent birds are obliged to hunt for food for them. They take large insects, especially dragonflies, in flight, and their phenomenal turn of speed and aerobatic skills enable them to hunt swallows, martins and even swifts.

The rarest bird of British heaths is the red-backed shrike. At one time it was quite widespread in many scrub habitats, nor was it confined just to heaths; but it has suffered the most dramatic decline of any bird. A century ago it bred throughout England and Wales as far north as the Scottish border; in the early 1950s it was still breeding on commons in suburban London; by 1980 it had declined to fewer than fifty pairs, and in 1988 just one pair was found in Breckland, bringing hordes of 'twitchers' to watch the two birds raise their chicks – fortunately within sight of a car park. This decline is

not fully understood, but is thought to be associated with climate and migration-pattern changes.

Several other, much more common birds help to make up the classic heathland avifauna, none of them confined to this habitat but all characteristic of it. Skylarks and meadow pipits are often the most frequent, especially where the heather is low and adjoins areas of grassy bog – sedge and cotton-grass tussocks provide good nest sites. Linnets may be quite numerous where the heath is dry, with clumps of gorse; gorse provides them with the ideal nest site, the very prickly spines affording in-built defence, and being evergreen, providing year-round shelter. Linnets, however, desert the heaths in winter, gathering in large flocks on arable farms and estuarine marshlands where food is easier to find.

Other common heathland birds are associated with scrub, and so both whitethroat and lesser whitethroat, and the grasshopper warbler, join the resident Dartford warbler in summer. Where heath adjoins birch-clad commons, willow warblers may also occur in some numbers. And stonechats, although more numerous on open stony hillsides, are still part of the characteristic heathland group of birds.

All heathland birds are declining rapidly because so much of their precious habitat is in southern England and therefore most pressurised by the human population. The greatest single threat comes from the urban developer.

Heathland Insects

Dragonflies are often the most conspicuous of heathland insects, provided there are bog pools in which they can lay their eggs. The water needs to be clean and unpolluted; many ponds on fragments of old heathland are often dirty and sterile. At good sites, like Thursley National Nature Reserve in Surrey or Studland National Nature Reserve in Dorset, many hundreds of dragonflies and their smaller relatives the damselflies may be seen in summer. The species list is impressive, and some of our rarest dragonflies occur only on heaths – for example, the brilliant emerald

dragonfly over the Ashdown Forest, and the white-faced dragonfly in Surrey. The golden-ringed dragonfly is one of the largest species in Britain, with a striking black and gold ringed body; it appears in mid-June and is most common on heathland. By mid-July as many as twenty species of dragonfly and damselfly may be seen around boggy pools at the best sites. The males patrol their territories and the females lay eggs in the pools by dipping their body in and out of the water, hovering over its surface.

A few butterflies are commonly found on heaths, including the grayling and silver-studded blue. The grayling is a member of the 'brown' butterflies or Satyridae, and appears in mid-July, flying until about early September. When disturbed it flies swiftly for a short distance, then lands on the ground where it instantly closes its wings, faces the sun and tilts over – apparently to avoid casting a shadow. The undersides of its wings are so appropriately coloured that it is very difficult to relocate it until it moves again. Its larvae feed on grasses. The silver-studded blue flies from mid-July to the end of August, feeding on heath flowers and laying its eggs on gorse, broom and heather. Several other butterflies occur regularly on heaths, including the green hairstreak, which in this habitat lays eggs on gorse – as opposed to rock rose or dogwood on chalk downs. Grassland butterflies such as small heath, meadow brown and gatekeeper also occur.

Among all the insects present on heathland, moths are one of its greatest delights, even though the variety of species is low because of the restricted habitat. The heather and gorse support such striking species as the emperor moth, whose large, well-camouflaged green and black larvae feed on heather and bramble and produce the beautiful large moths in April. The female emperor moth is noticeably larger than the male and on emergence from the pupa, produces a special scent from scent glands. This attracts the males

which may be seen flitting over the heather, tracking down the scent line. Experiments have shown that this scent is effective up to 1¼ miles (2km) away. The males fly strongly during the daytime, but the female is more nocturnal. This behaviour may be to provide some protection against daytime bird attack, and both sexes possess conspicuous eye-spots on their wings for the same purpose. The moths are large, and are the only British member of the silk moth family of tropical moths. The larvae spin large silken cocoons in the heather in which to pupate.

The pine hawk-moth is a large insect of pine woods, most prevalent in June and July, while other large moths found commonly on heaths include the fox-moth and the oak eggar.

There are many species of bee, wasp, beetle and spider which are characteristic of heathland; they are often brilliantly coloured. About 100 species of mining bees occur in Britain, often on sandy heaths where they dig burrows in the sand for their eggs. Also, there are dozens of species of wasp – sand, digger and other solitary wasps – which are common on heathland. The adults themselves feed on nectar, but they feed their larvae on live insects which are first paralysed and then placed in the nest cells. Some species hunt spiders, others eat caterpillars.

Beetles typical of the heathland habitat include some brilliantly coloured species such as the green tiger beetle; the larvae hide in sandy burrows and grab passing insects with their huge jaws. The wood tiger beetle is larger but less brilliantly coloured. Ground, carrion, dung and burying beetles may all be found on the heathland floor.

There are two types of grasshopper to be found on heaths: the rare heath grasshopper, restricted to just a few fragments of southern heath; and the large marsh grasshopper, which prefers boggy vegetation and heath pools.

Heathlands are rich in spiders, especially the genus Linyphia which is largely responsible for the millions of dew-filled webs of autumn. The bright pink crab spider, Thomisus, may be quite numerous on some southern heaths, hidden in heather flowers. It waits for unwary insects to land on the flowers, then seizes them from below.

Arctosa is a common wolf-spider – ants are its main diet – and it hides in a horizontal silk burrow

Green hairstreak butterflies appear in May and lay their eggs on heathland gorse. They have a wide variety of foodplants, from gorse to rock rose, bird's-foot trefoil, broom, heather, brambles and even garden peas and runner beans.

camouflaged with sand grains, darting out to seize the unsuspecting ants as they pass. It in turn is preyed upon by the hunting wasp Pampelius, which will paralyse the spider with its sting then lay its eggs on the spider's comatose body. Arctosa is superbly camouflaged, however, its colours merging perfectly with its environment – recently a melanotic black variety has been discovered on coal tips, having adapted genetically over the last few decades to blend with its man-made habitat.

Alternative hunting techniques are practised by other spiders – for example, the jumping spiders, Evarcha, are equipped with extra large eyes which are well adapted to their method of slowly stalking their prey, and finally leaping upon it. And the purse web spider, Atypus, which fashions a silk-lined hole a foot deep, capped by a web 'door' like a lid which it camouflages with bits of dried vegetation. The spider lurks beneath and when an unwary insect passes over its 'door', it impales it with its fangs through the web.

Heathland Reptiles

All six British reptiles – the adder, grass and smooth snakes, slow-worm, and sand and common lizards – are found on heathlands, especially those in the south. All six species hibernate through the cold British winter, reappearing in spring as the temperature rises. Most start their hibernation again in October.

Most reptiles are successful on heaths because these areas are dry and warm – this is why the adder, slow-worm and common lizard are also frequent on dry downland. Smooth and grass snakes on the other hand like access to boggy areas; indeed the grass snake is primarily a wetland animal. Heaths often possess real bogs, formed in places where the minerals have been leached by rain from

The adder is the most widespread British snake and particularly favours the warm, dry heathlands of the south and west. It is Britain's only poisonous snake, although its venom rarely kills humans. It hibernates from mid October to March and the young are born from eggs in August and September.

Common lizards emerge from hibernation in April and frequent dry heaths. They are inactive during the cool of the night, but their blood temperature rises with that of the air. They hasten the process by flattening their bodies on the ground to expose a larger surface area to the sun, until they are warm enough to begin to hunt. The young are born from June to August.

the sandy soil and have gathered much lower down as a hard, impermeable pan, preventing drainage and resulting in the accumulation of water in the soil layers above – the heathland bog.

The principal food item of both the adder and the smooth snake on this heathland habitat is the common lizard, abundant here and itself enjoying a rich food supply from the large number of insects and spiders which thrive in these dry areas.

The grass snake is not common in heathlands, but where there are bog pools there will be frogs and newts and these provide the snake with the main items of its diet.

Both sand and common lizards eat insects and spiders, as do slow-worms, and these will also take slugs and worms when available.

The adder is the most numerous and widespread of the three British snakes, and is found in many parts of the country including the Scottish mountains. It seems to prefer dry hillsides, heaths and commons, but as long as there are plenty of sunny places where it can bask it is fairly catholic in its choice. The best time to see it is in the spring when it emerges from hibernation. The males appear well before the females, and several may appear together on a warm March day. They like to bask on a sunny bank, especially early in the morning, but are surprisingly alert and extremely well camouflaged. Adders climb easily, and in places where they are numerous, especially on heathland, they may bask in the sun, lying across the tops of the heather scrub. However, like most reptiles, on the really hot days of summer they will retreat from the sun, and by midday will have slithered away into some shady nook or hole to cool down.

Adders eat lizards and often young mice and voles, and also worms and insects; they will also take birds' eggs and nestlings. As the adder is poisonous it pays to be able to identify it quickly. It is short and fat compared with the grass snake, rarely exceeding 2ft (60cm) in length, and is usually a dull, sandy yellow in colour with a distinctive broad, black, zig-zag line down the full length of the back. However, colour varies considerably, and there are black adders, just as there are black grass snakes.

The smooth snake looks a little like an adder, but does not have the adder's distinctive zig-zag line. It is usually a buff colour with dark spots along its body, and grows to about 2ft (60cm) in length. The smooth snake is very rare, found only on a few dry heaths in Dorset, the New Forest and in Surrey. Its habits in Britain are not really known because of its rarity, but basically it appears from hibernation in April, and feeds on small mammals and lizards; it will swallow larger prey whole, throwing a coil or two round it to hold it down.

The grass snake also favours the boggy areas on heathland, although it is even more numerous in fresh marshland and wet meadows. It is usually green and fawn with a distinctive yellow and black neck ring, and in comparison with the adder is long and slim, often exceeding 3ft (1m). Like the other snakes it feeds on live prey, and especially the frogs, toads, newts and tadpoles which abound in its marshy habitat.

All three British snakes mate in spring and the young – only a few inches (5–10cm) long – appear in early autumn, usually late August or September. Grass snakes lay eggs in a warm, concealed spot, often in piles of vegetation; smooth snakes give birth to live young which are enveloped in a transparent, membranous sac which quickly ruptures; and adders give birth to free-moving, live young.

Common lizards are usually abundant on heaths, and can be found on dry, grassy banks all over the country except in the outer isles – although they have been recorded at 3,000ft (1,000m) on Goat Fell in the Isle of Arran. They are small, often only 3in (7.5cm) long, and their base colour may be brown, yellow, red or green, with dark stripes both along and across the body. They are quick and graceful and can climb apparently smooth stone walls. Lizards are most often to be seen basking in the sunshine, often on objects which absorb the heat like logs, walls or large stones. They eat spiders and insects, including caterpillars and ants, and produce litters of some five to seven live young, often enveloped in a thin membrane, usually in July, although the whole breeding process may be delayed if they emerge late from hibernation because of a cold spring.

Sand lizards are found in the same restricted area as the smooth snake, on heaths in Surrey, Hampshire and Dorset, with the addition of one colony on the Lancashire dune coast. They grow to a larger size than the common lizard, and the mature male develops a bright green and pink coloration which is particularly brilliant in spring. They are surprisingly omnivorous, eating insects, spiders, worms, woodlice, beetles, centipedes and slugs. In captivity they have been known to lick up honey. Living in small, loose colonies, the sand lizard appears to be more socially inclined than its cousin the common lizard. Eggs are laid in June and July, each female producing six to fifteen eggs, and they hatch in August. All lizards return to hibernation again in October.

Slow-worms, often confused with snakes, are usually 12–15in (30–38cm) long, and a shiny grey or fawn in colour. The female has a black underside and in late August gives birth to identical, black-bellied, quicksilvered young, all enveloped in a transparent, membranous sac which ruptures as it passes out of the body. Slow-worms eat a variety of insects, and are especially partial to slugs. In turn they are eaten by adders, smooth snakes, hedgehogs, hawks and even pheasants. Like all lizards slow-worms can shed their tails as a defence mechanism if they are attacked. They are most likely to be found under logs and stones, but by late October will have retreated to hibernate.

CHAPTER 4

HEDGEROWS AND FARMLAND

A VAST area of the countryside is farmland and all of it is managed in one way or another by man for his own benefit. At least one third of Britain's earth surface is used for arable farming alone, without counting the grazing areas of sheep and cattle. It must be remembered that Britain's population is better fed than at any time in its history, and that farmers seek to make a living from food production in the modern world in the same way that others make a living from their jobs and careers. It is essential, however, for all to understand that farming and wildlife can, and must, co-exist. Farming cannot survive for long in an ecologically sterile environment, and it is time that the countryside was viewed neither as a food factory, by farmers solely concerned with maximising profit, nor as a museum for wildlife to which the well-fed townees, with profitable jobs, come for recreation. It must be an ever-changing, evolving environment providing both food and enjoyment, while at the same time maintaining conditions vital to our wildlife heritage.

Old-fashioned farming involved a labour-intensive industry which has declined as mechanisation has increased, to the point where in certain areas today prairie farming is the norm, with hedgerows and woods grubbed up indiscriminately to accommodate the larger, faster machines now essential to modern farming techniques. Ploughing, and the application of modern pesticides and fertilisers, all have a disastrous effect on plant and animal wildlife. In untreated meadows, flowers and grasses thrive – buttercups, daisies, yellow rattle, orchids and even fritillaries. And if the meadows are wet they may support rare wetland species like marsh gentian and small fleabane. However, when such ancient meadows are converted into grass leys by ploughing and planting with strong grasses, few of the original plants survive; they are unable to withstand the combined onslaught of ploughing, and the fertiliser and pesticide sprays.

Hedgerows between the fields themselves might well be considered miniature nature reserves, supporting a rich variety of birds, mammals, insects and flowers, and a recent estimate suggests that they cover some 450,000 acres (180,000ha) – a greater area than all the nature reserves in the British Isles put together. However, the total length of hedgerow – some half a million miles (800,000km) – has been cut by over 200,000 miles (300,000km) in the last thirty years. (These figures are from a recent Countryside Commission survey.)

Not only are hedgerows vitally important in their own right, they also link areas of woodland, acting as corridors of communication and easy passage for many plant and animal species which extensive arable farming might otherwise isolate, stitching woods and copses together with half a million miles (800,000km) of woodland edge. About a quarter of our hedgerows date from the eighteenth and nineteenth centuries when they were grown as boundaries after the Enclosure Acts of Parliament; some 200,000 miles (300,000km) of hedgerows were planted between 1750 and 1850. There was really very little destruction until after 1945 – wartime aerial photographs show an almost complete network of hedges, even in arable areas.

The oldest hedges divide the countryside into irregular, small field plots and date from Anglo-Saxon or even Roman times; they contain many species of shrub including hawthorn, blackthorn, holly, roses, hazel, and field maple. Many are actually remnants of old woodland, left in place as boundaries when the woods were cleared – hedges which contain woodland flowers like dog's mercury, bluebells and primroses originated in this way, as these flower species do not readily colonise hedgerows of their own accord.

Hedgerow management is essential if a thickset, bushy hedge is required. The traditional practice is to cut and tidy it by hand, but to lay a hedge like this is very labour intensive and is a method which has largely died out; modern hedging is usually carried out with a mechanical rip-cutter. If there is no system of management at all, the bushes will grow into straggly trees and large gaps will appear in the base.

Plant and animal interrelationships are very complex in hedgerows, simply because so many species are found in them. Hawthorn provides a good example, besides being the most common of all the shrubs found in a hedge: the sweet-smelling white flowers – called 'may-blossom' – are a source of nectar for many insects, while hawthorn leaves are eaten by over eighty different species of moth, like the brightly coloured vapourer, the lackey and the Chinese character moth. In summer, the spiny branches provide excellent nesting sites for many birds, from long-tailed tits and thrushes in the lower thickets, to jays, magpies and wood-pigeons in the higher branches. In the autumn its deep red berries provide a feast for many birds, especially the resident blackbirds and thrushes and the immigrant winter visitors – redwings, fieldfares and waxwings.

Hedges also play an important role as a substitute woodland edge, and so birds of this latter habitat are the most frequent; thus blackbirds and thrushes are common, as are wrens and robins, chaffinches and greenfinches, long-tailed tits and dunnocks. Other species use hedgerows as their main nesting site; for example 90 per cent of all common and red-legged partridges nest in the thick plant growth of the hedge base.

Both the structure and the plant species which make up the hedge are important in determining the bird population. Thus a hedge may contain trees of oak and ash, but unless some of these are old trees, it will support no breeding population of hole-nesters such as great tits, tree sparrows, jackdaws or stock doves. And unless the hedge is thick at the bottom, dunnocks, wrens and robins will also be absent as breeding birds. Hedges with a varied structure are best, so that birds can use different parts and levels of the hedge and its surrounds. Thus a chaffinch will sing from a hedgerow tree, nest in the dense thickets up against the trunk, and feed on the ground below the hedge and on the field verge nearby. The jackdaw will nest in a tree hole and feed on the nearby farm fields.

Autumn berries are eaten by flocks of thrushes, but they are also an important part of the diet of both wood mice and bank voles, the two most numerous rodents of the hedge. Competition is

PREVIOUS PAGE

An old hawthorn boundary hedge. The barriers they provide trap wind-blown seeds from other shrubs, as well as animal or bird-sown seeds and so develop a wonderful mixed habitat for insects, moths and a wide variety of birds and mammals.

RIGHT

Hedgerow shrubs provide a rich source of nuts and berries for wildlife. Rowan berries ripen in August and are immediately attacked by mistle thrushes, blackbirds and starlings. The rowan is also known as the mountain ash and is widespread in Britain.

neatly avoided, as the bank vole eats the soft flesh, while the wood mouse discards the flesh and eats the kernels.

The hedgehog's hunting ground is, of course, by no means restricted to hedgerows alone, but it is, nonetheless, a very common night-time visitor, snuffling its way through the hedge bottoms from April till November. Hedgehogs eat a wide variety of invertebrates such as worms, slugs and snails, and in this they compete with badgers and foxes. Daytime birds such as pheasants and partridges, blackbirds, thrushes and magpies also require a similar supply of invertebrate foods in good supply in hedgerows and on ploughed famland.

Butterflies are particularly noticeable members of the hedgerow community, and the first ones to emerge in spring will have hibernated through the winter as adults. True hedgerow butterflies may be identified as those whose food plants are most common in this habitat, even though they may occur elsewhere: for example, the orange-tip of late spring lays eggs on jack-by-the-hedge or hedge mustard, brimstones lay on buckthorns, while brown and black hairstreaks lay eggs on blackthorn, all classic hedgerow plants. The white-letter hairstreak suffered badly when its main food plant, the elm, was virtually wiped out by Dutch elm disease; however, it has managed to hold on in some areas by living on the growth of elm suckers which did not succumb to the ravages of the disease.

Between the hedgerows arable fields are totally artificial. Ploughed, sown, fertilised, sprayed and harvested to produce the end crop, they have a far lower value to wildlife than a similar-sized urban area. In such an environment, the intrusion of any wildlife species is regarded by the farmer as a nuisance, weeds or pests to be eliminated. Few of us would consider the skylark a pest species, yet despite the fact that, on average, there are only one or two pairs to the acre (0.4ha), some arable farmers do because it will occasionally eat the tops of sprouting beet and lettuces.

In days gone by, the accepted and usual way to improve soil fertility was to spread animal manure. Modern methods use fertilisers and trace elements, and a particular soil can be evaluated by a specialist laboratory very quickly and treated right away with the appropriate spray or artificial com-

Friesian cows on a Devon farm in spring. This ancient patchwork landscape supports a wide range of British wildlife, unlike modern prairie-style fields.

pound. Unfortunately some fertilisers are used indiscriminately, and this can have a dire effect on wildlife; for example, nitrates which leach out of the soil into watercourses, rivers and wetlands can cause rampant algal and weed growth, with disastrous ecological consequences.

The root of the problem is that in order to obtain the best possible return from a sowing, the farmer must control the quick-growing weeds which can cut crop yields by 20 to 30 per cent. In the modern world the solution is easy: the crop is simply sprayed with a suitable weedkiller. Similarly all the other pests and diseases – the rusts and mildews, flies or bugs – can be controlled with further herbicide sprays.

Whatever one's views about toxic chemicals, crop yields would undoubtedly be lower and less competitive if they were not used. Many farmers are now rather more conservation-conscious and try to maintain in a wild state the small areas of land which are less suited to agricultural development, but even these areas are still highly susceptible to the damaging effect of chemical sprays drifting from nearby fields, stubble fires, or irrigation schemes which can alter a habitat completely.

Some two hundred species of plant are considered to be arable farm weeds. These are usually annual plants which shed huge numbers of seeds into the soil, often producing ripe seed before the harvest, or ones which are small enough to escape farm machinery. The corn poppy is everyone's favourite and well adapted to arable land. It survives because its seed remains viable for as much as one hundred years, the plants springing up in great displays when undisturbed land is turned by the plough or dug up for roads or building sites. Seed production is prodigious, with over 1,000 seeds per capsule and possibly 20,000 per plant. Modern harvesting, however, often occurs before the seeds are shed, and although the corn poppy has survived, the same cannot be said of less hardy flowers, such as the corncockle, cornflower and

pheasant's eye, which are now virtually extinct. Only where a real effort has been made to retain them in experimental sites – like Butser Hill Iron Age Farm near Portsmouth – do they survive.

The plant and animal species of the natural world are linked together in a complex series of inter-dependent food chains, but on modern farmland these principles are often ignored. For example, some bird species which occur most often in farming areas and which are of considerable benefit to the farmer, are nevertheless hated by him. Rooks are major predators of leatherjackets in arable fields and eat many thousands of these destructive pests; a large rookery may therefore contribute greatly to an increased cereal crop in this way. However, by the time the grain has ripened in late summer the rook has changed its natural diet to autumn seeds – and this of course includes grain, now available in abundance. Therefore, the farmer, only too happy to enjoy the rooks' helping hand in spring, will do his best to drive them away in autumn and deprive them of their autumn food.

Many invertebrates play a vital role in maintaining soil fertility, the earthworm being one such; these break down the organic material in the soil into rich humus, and aerate the soil by their burrowings. Springtails are of great importance to soil fertility, feeding on soft vegetable matter and thus regarded as beneficial recycling agents. There would be many millions in just an acre of farmland, as long as it were not sprayed.

However, not all the inhabitants of the soil are harmless to agriculture – eelworms eat the roots of wheat, potatoes and many other crops, whilst wireworms – the larvae of click beetles – attack the roots of crops, especially winter wheat, severing the stems at ground level. Whole fields may suddenly wilt if there is a large wireworm population.

However harshly the modern farmer may be criticised for his methods and the chemical applications he uses, if he is conservation-minded and well disposed he can provide a great variety of mini-habitats for wildlife about his farm. Conservationists might find it more rewarding to encourage rather than condemn the modern farmer out of hand.

101

Hedgerow Flowers and Insects

Which particular flowers grow in a hedge depends very much upon its origin. If it contains primroses, dog's mercury, bluebells and yellow archangel, or a similar mixture of the flowers of ancient woodland – all very slow colonisers of new ground – then it is likely that the hedge itself was part of the original wood, left to mark a boundary when the wood was cleared. Such a hedge will probably be quite old and its antiquity will be further confirmed if it contains a number of shrub species including hazel, dogwood or field maple, all of which are slow to spread to new hedges.

Sometimes the most striking feature of a hedgerow is its climbers, which use the hedge for support in the same way as they do the trees and bushes along the edge of a wood. Roses are found in many British hedges, and there are several spec   burnet rose appears first in May, followe   eet-briar and field rose in Jun   northern spe   nate because   ig to the h   where in th   t bounti-f   ians and   ble have   ld regard

                              r support
                              ir flowers
                              ollination
                              nbers in-
                              eysuckle,
                              st often in
                              e and rose

will be in bloom together. The rose is pollinated by a bee or a wasp, but what pollinates the honeysuckle with its long corolla tubes? Most flowers with hard-to-get-at sources of nectar are pollinated by moths, and especially by hawk-moths with their extra-long tongues, the moth hovering in front of the flower and probing deep into the corolla to lap up the nectar. Red honeysuckle berries, poisonous to humans but eaten by some birds, appear in autumn.

Black bryony also produces poisonous berries, which hang in long scarlet sashes in autumn; the berries contain a toxic glucoside, saponin, and even birds avoid them. Black bryony favours the shaded, humid side of the hedge; whereas white bryony, quite unrelated, is a vigorous climber of the drier, sunny aspect and is found on calcareous or base-rich soils. It clings by tendrils which are so sensitive that if touched, they will spring into a coil. This plant also produces poisonous red berries.

On chalk and limestone soils the most notable climber is traveller's joy, or old man's beard, its white, feathery, plumed seed heads often filling a chalk-down hedge in late autumn. Ivy is another extremely important hedgerow plant, providing nectar and pollen for insects in late autumn and ripe blue berries for birds the following spring. Under its insulating layer of evergreen leaves it also provides hibernating sites for hedgerow insects, and shelter for small birds at night.

Other hedgerow climbers are really scramblers, the plant stems threading their way towards the light. Woody nightshade is a fairly common hedgerow plant with clusters of blue and yellow flowers in early summer and bright red, poisonous berries in autumn. All parts of the plant taste 'bitter sweet' – its other common name – because they contain a strongly poisonous alkaloid, Solanine. Other scramblers include tufted vetch, goosegrass, and in the south wild madder.

Among the two or three hundred traditional hedgerow flowers there are many which prefer disturbed ground – the willow-herbs, yarrow and shepherd's purse, the flowers of arable farmland. These are the ones regarded as 'weeds' by the farmer; however, they do in fact play host to countless insects of a great variety which help pollinate his crops, and eat the pest species which may harm them. Perhaps a compromise could be reached,

103

the farmer tolerating at least some of the hedgerow and its edge.

In early summer the hedges are filled with umbellifers, cow parsley, alexanders, fennel, wild carrot, hogweed and many more. These grow tall, and are alive with insects; while in the hedgerow base grows hedge mustard – among many other things, but this is notable as the food plant of that classic hedgerow butterfly, the orange tip. Other common hedgerow flowers which thrive in the half-shade conditions include barren and wild strawberries, bush vetch, greater stitchwort, hedge woundwort,

herb bennet, herb Robert, wild arum, sweet violet, primrose and lesser celandine.

The growing leaves of the wide variety of hedgerow plants provide food for the huge numbers of insects associated with them. The stinging nettle alone supports twenty-seven insect species, and like the bright green nettle weevil, these feed on nothing else; a further seventeen use the nettle as one of their major sources of food.

Butterflies will be most active in sunshine, with twenty or more species along a good hedge: brimstones, feeding on alder buckthorn, followed

LEFT

An orange tip butterfly feeding from hawthorn blossom in early May. Its principal food plants are cuckoo-flower and garlic mustard – commonly known as Jack-by-the-hedge. The female lacks the distinctive orange tips of the male. The underwings of both sexes give an appearance of green mottling on white which provides one of the best examples of camouflage in butterflies when resting on the flowerhead of garlic mustard.

RIGHT

The greater periwinkle is just one of the many hedgerow flowers which develop a climbing or scrambling habit. It flowers in spring between February and May.

by orange tips on hedge mustard, meadow browns, gatekeepers, skippers and ringlets on grasses, small tortoiseshells and peacocks on stinging nettles, large and green-veined white on various cabbage family plants, and brown hairstreaks on blackthorn – all these are just a part of a substantial butterfly fauna.

Hawk-moths use hedges as well; the privet hawk is found almost exclusively in privet hedges, and the elephant hawk feeds on willow-herbs, especially the great hairy willow-herb found in many damp ditches.

It should not take long to realise that there are many more members of the bee family than just the honey bee and the bumblebee that frequent the hedgerows. In fact there are about 250 different kinds of bee in Britain, including 18 different bumblebees. Bees and wasps are brightly coloured to warn off predators, and several other creatures mimic these warning colours, thereby gaining a measure of protection; for example, some hover-flies imitate several different bumblebee species, while wasp beetles are mimics of the common wasp.

Birds of Hedgerow and Farmland

Only a few birds breed regularly on the open fields – these include lapwings, skylarks, meadow pipits, and common and red-legged partridges. Both the lapwing and the partridge have declined, however, the lapwing because cereal fields are rolled in early spring, just as it has laid its eggs in the young crop; the partridge primarily because of the intensive use of insecticides, which reduces the food supply for the chicks to a critically low level.

The Common Bird Census of 1986 showed densities of 1.6 pairs of partridges per sq km in Dorset, to 7.2 pairs on a Suffolk arable farm, with an average of 2.76 pairs, giving a possible British population of 400,000 pairs, but declining. The red-legged partridge, first introduced in 1673, shows some 100,000 pairs. By the same methods the British Trust for Ornithology estimated the lap-

wing population to be 200,000 pairs in 1976, but there has been a serious decline since.

Skylarks, at five to ten per sq km are one of Britain's most widespread birds, with perhaps 2 million pairs. In winter both lapwing and skylark populations gain vastly from influxes of continental birds, with up to 1 million lapwings and 25 million skylarks present.

The vast network of hedgerows which covers most – but now, alas, not all – of lowland Britain, is inhabited by many woodland species. It is most valuable to birds like thrushes and finches which are largely wood-edge birds, and which need trees and bushes for nesting and for song posts and cover. Species that breed and thrive in hedges are those that habitually feed out in the fields or along the wood edge; birds which find their food wholly within the wood, such as nightingales and marsh and willow tits, are rarely found breeding in hedgerows.

Hedges thus provide the basic conditions necessary for many common wood-edge species, although they are relatively unimportant for the rarer ones. However, very large numbers of our common woodland birds would not exist on farmland were it not for these hedges, mainly because they provide a wonderful variety of food. There is nearly always a rich supply of many insects, fruit and seeds – the more species of shrub and flower in the hedgerow, the greater the number of different insect species that will occur. Similarly, a great variety of edible fruits and nuts, such as rosehips, haws, sloes and hazelnuts, will be found in an old hedge with many types of shrub.

There will be berries for fieldfares and mistle-thrushes, snails for song-thrushes, worms for blackbirds, ash seeds for bullfinches and grass seeds for linnets, thistles for goldfinches, mice for owls, and sparrows for sparrowhawks. In winter, all five British thrushes, the blackbird, song- and mistle-thrushes, fieldfare and redwing will probably depend upon the hedgerow berries, especially if the fields are frozen. If winter conditions are good, then they take a varied diet of invertebrates from the fields and berries from the hedgerow. (In turn the thrushes take the seeds with them, releasing them through their droppings, thus dispersing the berry-bearing bushes of the hedge itself.)

Bill size and shape are determined by food pref-

The lapwing population has fallen considerably in the last ten years due to changes in farming practice. Its tumbling erratic display flight and its distinctive call – a shrill 'pee-wit' – make identification of this species very simple. Highly gregarious, flocks of lapwings descend on farmland in autumn and winter, breaking up in late February or early March for mating.

erences, finches having stout bills to crack seeds, while dunnocks, or hedge sparrows, have long, thin bills to extract insects from crevices. But research has shown that bird-feeding ecology is not that simple, and that different food sources are exploited as they become most abundant at various times of the year. Thus dunnocks eat insects in summer but turn almost entirely to small seeds in winter; seed-cracking finches are not always veg-

etarian, but feed a high-protein diet of insect larvae to their chicks in summer – and in early spring, when seeds are scarce, they often turn to tree buds.

In spring the resident hedgerow birds are joined by summer visitors from Africa, although only a few of the many arrivals are commonly found breeding in hedges. The whitethroat is a classic 'scrub' warbler, closely related to several similar species found abundantly in southern Europe. It uses the tops of bushes to launch into its distinctive song flight before 'parachuting' out of sight into the base of the hedge, and nests a foot or two (half a metre) off the ground in the thicker, lower level tangles of bramble, briar and nettles. The lesser whitethroat is another hedgerow scrub warbler. Otherwise, the willow warbler and spotted flycatcher are the only other summer visitors

to breed regularly in hedges, birds such as chiff-chaff, blackcap, redstart and pied flycatcher preferring real woodland.

Different hedgerow styles provide differing breeding sites for birds and thus support differing bird communities. For example, a clipped, low, dense hedge will hide many small nests but have no song posts for dramatic singers to proclaim their territory, nor tree holes for hole-nesters. On the other hand, if it is old and straggly, with interspersed large trees but no real base, it will be more suitable for the mistle-thrush, magpie, wood-pigeon, collared dove and starling, blue and great tit, tree sparrow and stock dove in the tree holes. Occasionally one of the owls – little, tawny or barn – will use a larger hole in one of the older trees. But with no base vegetation there will be no small scrub birds.

From the birds' point of view the ideal hedge has a mixture of standard trees, and some tall and overgrown thorns with spreading bushy base, over a damp ditch filled with nettles, cow parsley and wildflowers. So as well as the birds of tree top and tree hole, this ideal hedge will be able to shelter

The common or grey partridge has declined because toxic chemicals kill its insect food, and fields ploughed right up to the hedges deny them their marginal grassland. It feeds mainly on vegetable matter, seed, grain and some insects. The young partridge is dependent on caterpillars as food during its early development and is therefore susceptible to cold late springs.

and support the more numerous hedgerow birds of dense scrub – blackbirds, robins, wrens, long-tailed tits, song-thrushes, dunnocks, chaffinches, linnets, redpolls, bullfinches, greenfinches, yellowhammers and corn buntings. If the ditch is wet there may even be a substantial population of moorhens. Pheasants and partridges may well use the hedge base, feeding on the surrounding farm fields.

Winter brings flocks of gulls, lapwings and golden plovers to the farm fields, pied wagtails to roost in the warmth of the greenhouses, finches to collect the gleanings from the harvest, and hordes of thrushes from the Continent to eat the hedgerow berries.

108

Barn Owls

The beautiful barn owl, as a hunter always of invaluable service to man, keeping the rodent population in check, is a bird of rough grasslands, marshes and hedgerows. Once fairly numerous, it has become a rarity in Britain, and though the reasons are not entirely clear, the pressures exerted by the modern world have obviously contributed to its disastrous decline.

More intensive farming, coupled with the vastly increased use of toxic chemicals on farms, has significantly reduced the barn owl's food supply, sometimes poisoning the owls themselves. No longer are there extensive lowland areas of rough pasture sheltering innumerable grass-eating field voles. Huge areas have been converted to insecticide-covered and treeless wheat prairies, and this means that in some places the owls have gone for good. Ancient hedgerow trees full of holes and undisturbed farm buildings used to provide the barn owl with its best breeding sites; hedgerow destruction and the conversion of old barns into luxury homes have also therefore contributed in large part to the owl's demise.

Contrary to popular myth the barn owl does not actively seek the companionship of humans; it just so happens that its prime habitat is rough grassland, and this used to coincide with British farmland. An examination of nest record cards held by the British Trust for Ornithology shows that the owl's favourite nesting sites were tree holes – 42 per cent; then in barns, 23 per cent; and in haystacks, 7 per cent.

Barn owls are easily disturbed, both at their nest site and at daytime roosts, and at the latter may desert the site altogether if disturbance continues. The increase in high-speed road traffic, and especially the enormously extended mileage of motorways and trunk roads, is a major cause of death for these slow-flying hunters, particularly as some of them have taken to hunting the relatively undisturbed motorway verges at night, replacing the more familiar daytime kestrels. A staggering 5,000 owls are thought to die on the roads every year, fully 25 per cent of one year's production of nestlings.

Another influence on population numbers is hard weather; the barn owl is very susceptible to the cold, frozen conditions of a hard winter, when it may be unable to obtain its prey from the frozen, snow-covered ground.

The current barn owl population is at a low ebb in most parts of the country. A 1986 estimate (from a survey carried out by the Hawk Trust over the period 1982–6) suggested that there were only 5,000 pairs, a decrease from some 12,000 pairs estimated in the previous survey in 1932. The barn owl survives in reasonable numbers only in southern Scotland, where farming depends on dairy and livestock production, and the pattern of farmland with small hedged fields has remained relatively unchanged. Until recently, East Anglia supported a good population, but has lost many owls over the past five years. The reasons for this are almost certainly associated with hedgerow removal, the high use of insecticides on wheat and rape fields, and the popular and increasing trend of converting old barns to houses. An indication of how the barn owl has suffered in areas of industrial expansion is shown by an extensive survey in Surrey in 1986/7, which showed only fifteen pairs in the whole county. The barn owl is widely distributed throughout the world, but this gives a misleading impression of abundance, as in many countries it has suffered huge population losses for many of the same reasons as in Britain.

The barn owl stands about 13in (33cm) tall, and is some 2in (5cm) smaller than the much more numerous tawny owl. In Britain it is usually white underneath with pale, buff-flecked upper parts, giving it its famous ghostly appearance, especially if it happens to waft by on silent wings in the dusk. When it calls, which is not very often, it produces a high-pitched shriek – this has led to its other name of 'screech owl' and has served to augment the superstitions associated with it.

Like all owls, its ears are set asymmetrically on its head so that it can pinpoint the exact position of its prey in the grass as it plunges down feet first to grasp it. Its flight is particularly silent even in the quiet of night, because its feathers are specially adapted with a multitude of fine filaments which act as silencers. Each individual owl has its own particular hunting technique, although they all employ the 'flap and glide' method at times. Hovering is a regular tactic, for some much more than

others, and in cold weather the 'beat-up' method may be employed, whereby the owl flaps its wings vigorously over sparrows and starlings roosting in the bushes. One of their most important and energy-saving techniques is to 'post hop'; this entails scanning the ground intently for several minutes from a fence post and if no food appears, flying to the next post and repeating the performance.

Courtship begins late in February and eggs are laid from March onwards. If the food supply for a particular pair is good there may be a second brood in late summer, but this appears to be much less frequent than was once thought and many pairs seen with late broods may not have bred earlier in the year. Four or five white eggs are laid, and the same nest site may be used for thirty years or more. Incubation lasts for thirty-two to thirty-four days, and once the chicks hatch the parent owls become even more regular in their habits than normal; they will leave the nest site about forty-five minutes before sunset – if the food supply is good they may not travel far, returning every fifteen to thirty minutes with a mouse, vole or rat. Young barn owls take a long time to fledge and are in the nest for nearly three months, but once fledged, the whole brood lives with its parents for a little while until the young owls are driven away by the adults

110

The barn owl is fast becoming an endangered species in modern Britain. Watching a barn owl hunting in the dusk you will be struck by its silent, ghostly appearance and its long moth-like wings, designed for gliding and slow quartering of the ground.

in late autumn. It is hardly surprising, therefore, that the largest number of road casualties occurs in the autumn and early winter, since this is when the inexperienced young owls are wandering further afield in search of a territory of their own.

In cold weather the owls often hunt within the confines of the barn, and this is something this particular species must have done for several centuries – but the old-fashioned, rodent-filled barns have given way to clean, modern buildings in which the rodents are controlled with poison bait. The food supply is lower, and the owl is at high risk of picking up a dose of poison from a dying mouse or rat.

Research on the diet of the barn owl is easy, because the undigested skulls and fur of the prey animals are regurgitated in pellets – themselves a good clue to the presence of owls, since they often accumulate in piles under roost sites, particularly obvious in barns. Rodents form 85 per cent of an owl's prey, though which species depends on the hunting area; however, field voles, rats and mice are the most frequent, followed by shrews and bank voles. Small birds such as sparrows are occasionally taken by some individual owls. In a year, a pair of owls will take an average of 4,500 prey animals each weighing 1oz (28g): a vole provides one 1oz (28g) unit and an adult rat may provide five units.

It is illegal under Schedule 1 of the Bird Protection Acts to disturb barn owls at or near their nest sites. Farmers could help to assure the barn owl's future, too, by putting nest boxes in good barn sites. All in all, the barn owl is severely threatened, and conservation measures are urgently needed to prevent the total demise of this useful, charismatic bird of the night.

Rabbits and Hares

There are three species of rabbit and hare in Britain – the well-known and ubiquitous rabbit, the brown hare and the blue or mountain hare. The rabbit was brought to Britain by the Normans for fur and meat, and it is possible that the brown hare was a Roman import; all three belong to the mammal order Lagomorpha. Lagomorphs have six incisor teeth and are therefore different from rodents, which have only four. There are some forty species worldwide.

The rabbit originated from Spain; there is no mention of it in Domesday (1086), and it first appears in written records in 1176. During Norman times it was called a 'coney'; the name 'rabbit' is from the fourteenth-century French word *rabbete*, and originally applied only to the young. By 1950, 40 million rabbits were killed annually in Britain without denting the population. Densities of forty per acre (0.4ha) could be found in good areas.

In 1953 someone unofficially, but quite deliberately, introduced myxomatosis into Britain and this killed 99 per cent of all the rabbits within two years. The rabbit was rare until the 1960s, but some of those which survived had gained a genetic immunity to the disease, and by 1988 the population had reached 20 per cent of the pre-myxomatosis levels, and much higher in some areas.

Rabbits are found throughout mainland Britain and Ireland and on all but the most isolated islands. Brown hares do not occur in much of the Scottish Highlands, nor naturally in Ireland, although some have been released there, but they are found in the rest of lowland Britain. Mountain hares are found throughout Ireland and in the northern upland districts of England and Scotland from the Peak District northwards. The Peak and Pennine hares have also been introduced.

Rabbits are generally easier to approach than hares, although both are very alert at all times because they feature on the menu of many predatory animals and birds. They have acute hearing as well as a keen sense of smell and good eyesight. Rabbits are easily located by way of freshly dug holes, their droppings and the paths they make through the

vegetation. An occupied warren is usually obvious, although a long line of holes sheltered by a hedge is less clear. A warren is often on a slope such as the downs, amongst sand dunes or in a railway bank, because the drainage is better.

Rabbits like areas of short grass for their young, which do not then get as wet as they would in damp vegetation; they also like clumps of dense cover close by, into which they can bolt at the first sign of trouble. The hare prefers wide open spaces and long grass for its 'form' – a form is a depression in the ground which serves as a temporary home. Hares are difficult to approach closely, and are best watched from a distance, perhaps from a hedgerow across wide farm fields. Mountain hares are best seen in March or April when the snow has gone but the hares are still white; they are most likely to be found along rocky hillsides, out of the wind and below the snow fields.

Rabbits and hares are vegetarian, eating large

quantities of greenstuff, mainly grass. However, their digestive system is organised so that they can spend most of the day in cover, digesting the food which they have gathered hastily in quick feeding forays. This involves a process known as 'refection', whereby a dark, mucus-covered faecal pellet containing only part-digested plant food is produced, which the rabbit or hare takes straight from the anus and swallows. Food is thus digested twice, once to break down the cellulose and then to extract and absorb the nutrients.

The gregarious nature of the rabbit results in distinctly social habits; this also provides as many ears as possible for defence purposes. They live in groups of up to eight or ten, each group defending its own small territory from the several others that inhabit the warren. In spring the doe, or female rabbit, will dig her breeding tunnel, usually about 3ft (1m) long, and after a twenty-eight-day gestation period, four or five blind and helpless kittens are born into a grass- and fur-lined chamber in the tunnel. The breeding season lasts from the end of January to July, and there may be four to six litters. However, survival rate is low, only about 10 per cent; but when populations do get too high, rabbits have an efficient system of birth control whereby embryos die at an early stage of development and are absorbed back into the uterus. This is known as 'resorption'.

At birth the kittens weigh about 1oz (30g); adults vary between 3 and 4½lb (1.4–2kg). The young are normally fed at dusk, inside the burrow when they are very small, and at the burrow entrance as they grow larger. The doe will suckle them for only three or four minutes and only once a day, a system of minimal care which is thought to be a defence against discovery by predators. Nevertheless, baby rabbits grow quickly and are sexually mature at three to four months old.

Baby hares are called leverets and initially their upbringing has certain similarities to that of the rabbit. As soon as they are strong enough they will move away, individually, from the birthplace and spend the day crouched in a separate hollow in long grass. They, too, are fed only once a day at dusk and only for a few minutes, the mother standing upright to watch for danger in the open field. However, at birth, baby hares are quite different from baby rabbits: they are born fully furred, and in full possession of all their senses – eyes, ears and nose; and after the first few unsteady minutes they can run, too.

Hares are noticeably larger than rabbits, the average weight for an adult being nearly 8lb (3.6kg) for the brown hare and 7lb (3.2kg) for the mountain hare. They produce fewer young than rabbits, with a gestation period of some forty-four days in the brown hare, and fifty in the mountain hare, with perhaps three litters of two to three babies which take some eight to nine months to grow to maturity; thus they do not breed themselves until the following year.

Because of their tall black-tipped ears and long, powerful back legs hares are easily distinguished from rabbits. In addition, the mountain hare has an all-white tail which distinguishes it from the brown hare. However, it still keeps the black-tipped ears even in winter when it moults into its white fur

LEFT
Rabbits are sociable animals living in large communities divided into groups led by a dominant buck. Throughout the year the life of a rabbit centres on a network of burrows, the warren, which forms its underground home.

RIGHT
Brown hares are slowly declining in numbers, possibly due to the lack of long grass cover on the perimeters of modern farm fields. At birth, leverets are in full possession of all their senses and are covered in fur, adaptations typical of a 'plains' animal which is likely to be chased by predators while still very young, without the safety of a bolt hole to rely on.

coat; this winter coat obviously provides excellent camouflage in the snow-covered hills. In its summer coat it is more easily confused with the brown hare, but one distinguishing feature is that the black-topped tail of the brown hare is usually turned down as it runs away.

Brown hares are most numerous on open farmland with areas of long grass for cover. Mountain hares occur in largest numbers on northern grouse moors, where the land management which produces new heather shoots for the birds also produces new shoots for the hares; heather is their staple food. Brown hares are surprisingly common on airfields, including a now-isolated population at Heathrow. They like the long grass and the space, and take particular delight in competing with the planes, rushing alongside a taxi-ing aircraft.

Small Mammals of Hedgerow and Farmland

Nearly all the small mammal species found in Britain are represented in fields and hedgerows, although many of course, are common to other habitats, too. The most numerous species are the common and pygmy shrews, the field and bank voles, the wood mouse and harvest mouse.

The common and pygmy shrews are insectivores, not rodents, and are related to hedgehogs; they have mobile, pointed snouts, and tiny eyes, small ears and a medium-sized tail, and thick, dark-brown fur – but their overall size is tiny, half the size of an adult wood mouse. Voles have flattened, blunt faces, small ears and eyes and short tails, while mice have much more pointed faces, with

Harvest mice are animals of the 'stalk zone' during the summer, in cereal fields, ditches and reeds. In winter they live largely on the ground. The mouse's breeding nest is the size of a cricket ball and is slung between two stalks.

larger eyes and ears and tails longer than their bodies.

Field voles and shrews are essentially grassland animals, although shrews may also be abundant in dense undergrowth where this is mixed with long grass, along the hedgerow edge and in a scrub habitat. Field voles live the whole of their lives in grassland, though rarely on the surface – more usually they fashion a network of tunnels through the lowest stems and grass roots. They eat grass and sleep in tennis-ball-sized grass nests buried in the roots of large tussocks. Their population is cyclical, rising to a notable peak every three or four years when densities of several hundred per acre (0.4ha) may occur. These 'plagues' may cover wide areas, or they may be localised, but they certainly influence the breeding and concentration of predators such as kestrels, barn and short-eared owls, hen harriers and buzzards, all of which may produce larger broods as a result.

The field vole's weight varies with the stage of the population cycle, averaging 1oz (25–30g), but increasing to 1¼oz (35g) when food is abundant. Research shows that their activity is spread over the whole twenty-four hours, but intensifies at night. The home range is small, few animals moving more than 30yd (29m) throughout their lives. The gestation period is twenty-one days, with three to six young in each litter; they are weaned at fourteen to eighteen days, and are sexually mature at twenty-one to twenty-five days – thus allowing a rapid rise in the population if conditions are suitable. Voles and mice live, on average, for fifteen months, a little longer than shrews.

A shrew's weight varies from season to season, but on average a common shrew weighs about ⅓oz (10.5g) in summer and less than ¼oz (6–7g) in winter. Pygmy shrews vary from ⅒ to ⅕oz (3 to 5g) – or less than a fifth the weight of a house mouse. All shrews are very pugnacious, squabbling with any other shrew they encounter. They rarely wander more than 50–75yd (48–70m).

The shrew's main predators appear to be birds, especially owls, which swallow them whole. Mammalian predators such as foxes, stoats, weasels and domestic cats certainly kill many shrews, but they appear to taste awful when chewed and are often discarded. Shrews have flank musk glands which are presumably used for territory marking in their

complex tunnel systems in the root zone, and it may be these which make them unpalatable. They also have a toxic saliva which is used in subduing their insect prey.

Both the wood mouse and the bank vole are primarily woodland animals, and both are frequently resident in the hedgerow. However, live trapping with Longworth traps has shown that wood mice not only inhabit the hedge but are found all over the neighbouring fields, especially cereal fields, foraging for food, unlike the bank vole which does not stray more than a yard from the hedge base.

Much rarer than the shrew, wood mouse and bank vole is the delightful harvest mouse, the traditionally heralded mouse of the farm fields. It is on the edge of its range in Britain, and is most often found in the south and east. It is very distinctive – tiny, no bigger than a 10p piece, with a long, truly prehensile tail, used constantly for support when climbing. The adult is russet in colour, almost orange on the back and flanks, and there is a vague dorsal band of dark-tipped hair along its back. Its undersides are white, making it a most appealing little animal.

Recent research has shown that although increased mechanisation of farming and the use of pesticides have no doubt affected the general population level, this is not to the degree previously imagined. The harvest mouse still thrives in ditches, hedges and tall vegetation and is often to be found in considerable numbers in reed beds or reed-filled ditches along cereal fields. In summer it moves out into the cereal fields themselves, and spends most of its time climbing, building nests a foot or two (30–60cm) up in the 'stalk' zone in which it lives. In late autumn it retreats to ground level, taking up residence in the warmer, sheltered micro-climate of grass tussocks and roots. Deep snow does not cause either the mice or many of the other small mammals any bother, as it actually insulates the ground and root layer very effectively from cold winds.

The average litter size of the harvest mouse is five, but life expectancy is only about a year, so most females produce no more than one or two litters. The harvest mouse, although very small, weighs on average about ¼oz (some 6 or 7g) – 50 per cent more than a pygmy shrew.

Hedgerow Hunters: Foxes, Stoats and Weasels

There are far fewer foxes in a given area of the countryside than there are in a similar area in the suburbs of British cities. This is because towns provide more food in a smaller area, enabling individual animals to make a living in a smaller 'home range', thus resulting in a greater density of population overall. Out in the country foxes range over a much wider area and on open hills and moorland the home territory may be as large as 2,500 acres (1,000ha), or 4sq miles (10sq km). In lowland farmland their range is more restricted, on average about 500 acres (200ha).

The fox is a medium-sized animal, the average weight of the male being 15lb (6.8kg) and the female 12lb (5.5kg); its average length including the tail is 3½ft (105cm). Contrary to popular belief, there is absolutely no physical difference bet-ween country and city foxes, and some actually commute from the countryside into the towns and suburbs to obtain food at night. Similarly, town foxes are as well fed and sleek as their country cousins. Foxes which look half bald in summer are usually vixens with cubs, probably going through their summer moult.

Diet is one of the main differences between city and country foxes. Country foxes are hunters to a much greater extent than those in the city which obtain most of their food by scavenging. Much research has been carried out into the feeding habits of the fox, which is an opportunist, taking food as it appears. In the countryside the bulk of the diet is small mammals: rabbits, hares, rats, mice and voles. Some foxes have clear preferences, and they will all change their diet to take advantage of any seasonal abundance of food. Rabbits and field voles are the preferred prey – assuming a good supply of rabbits, since myxomatosis still affects the wild rabbit population. The speedy, brown hare of the lowland does not often feature as fox food, but mountain hare does regularly in the diet of foxes in the Highlands. In Ireland, where there are no field voles, brown rats are the most common

The weasel is small enough to enter mouse holes. It is distinguishable from the stoat by being smaller and without a black tip to its tail. Unfortunately, like the stoat, it is still persecuted by gamekeepers.

food item after rabbits and hares.

Birds are surprisingly infrequent as fox food, although ground-nesting species, like ducks, waders, pheasants, gulls and terns, are obviously extremely vulnerable. Intensively reared game birds with artificially high population densities are always at risk, especially since cubs require food at precisely the time of year when these birds are nesting. Bird colonies, such as gulls and terns, can also suffer badly. One black-headed gullery of 8,000 pairs lost 5 per cent of the adult gulls, or 400 birds, to just four foxes living nearby. Juvenile gulls suffered even more, being unable to fly; they made up 28 per cent of these foxes' diet, and were the most important food item after rabbits.

After it has rained, earthworms are frequent items of the fox's diet, and various fruits feature largely in autumn. Foxes are also partial to blackberries, reaching up on their hind legs to delicately pluck the ripe fruit. Even in the countryside a lot of food is obtained by scavenging, anything freshly dead being promptly eaten.

The breeding rut is from December to February, with cubs born from January to April, after a gestation period of fifty-two days; the average litter size is 4.7. The young weigh some 3½oz (100g) at birth and reach adult size at twenty-six weeks. Fox cubs are usually above ground for an hour or so before sunset, and will emerge much earlier in the day where disturbance is minimal. They may stay in the earth until June, when the vixen moves them away. Cubs are inexperienced and may be watched from just a few yards away as long as you are upwind of the earth. The vixen, however, is unlikely to be fooled and will often make a complete circle of the area before bringing food in to the cubs.

The other two common hunters of hedgerow and farmland are the stoat and the weasel, but they are much smaller – even a large stoat weighs under 1lb (0.45kg). Yet these two tiny hunters have a reputation for ferocity out of all proportion to their size. Both are slim, lithe and fast – the weasel is not much bigger than a large mouse but its body is comparatively elongated; the stoat, with a black tip to its tail, is noticeably larger. Its main prey is the rabbit, and there seems little doubt that the stoat population fell dramatically when the rabbits vanished with myxomatosis in the 1950s. Only now, some thirty-five years later, is it returning to its previous numbers.

Stoats and weasels are truly carnivorous, rarely eating anything but other mammals, mainly rodents. Both kill by biting the back of the neck and hanging on tight, and both can deal with prey twice as large as themselves, the stoat killing rabbits and the weasel sometimes rats; usually, however, they tackle the smaller voles and mice and are thus useful predators in the countryside. Both eat about 30 per cent of their body weight daily which, for a large male weasel amounts to about one and a half voles, and for a stoat a large rat. They also kill nestlings and will take young pheasant chicks, and are therefore constantly persecuted by gamekeepers and the shooting fraternity throughout the country as 'vermin', which they clearly are not.

Both stoats and weasels are active throughout the day and night; they must have food at frequent intervals for two main reasons – their slender bodies have a large surface area and lose heat rapidly; and they have a high metabolic rate. However, their body shape enables them to follow their prey into runs, stoats going down rabbit burrows and weasels along the narrow runs dug by voles and mice.

They are rather solitary animals, although stoats may hunt in pairs. Those country tales which record 'gangs' of them together are in fact describing a family group, the female followed by her litter which may number as many as eight kittens. Generally the males have large territories, and two or three female territories may lie within the range of one male.

Stoats and weasels are likely to be seen only by chance, yet they are very inquisitive which is often their downfall – once they have found some cover, they may well sit up on their hind legs to look about them. If you make high-pitched squeaks with pursed lips they can sometimes be fooled into approaching quite closely. Try not to handle a wild one – if you do, you will surely regret it!

The fox, small enough to go unnoticed and highly mobile, has adapted to man's towns and cities with ease during the last fifty years. Its omnivorous diet has helped this urbanisation greatly.

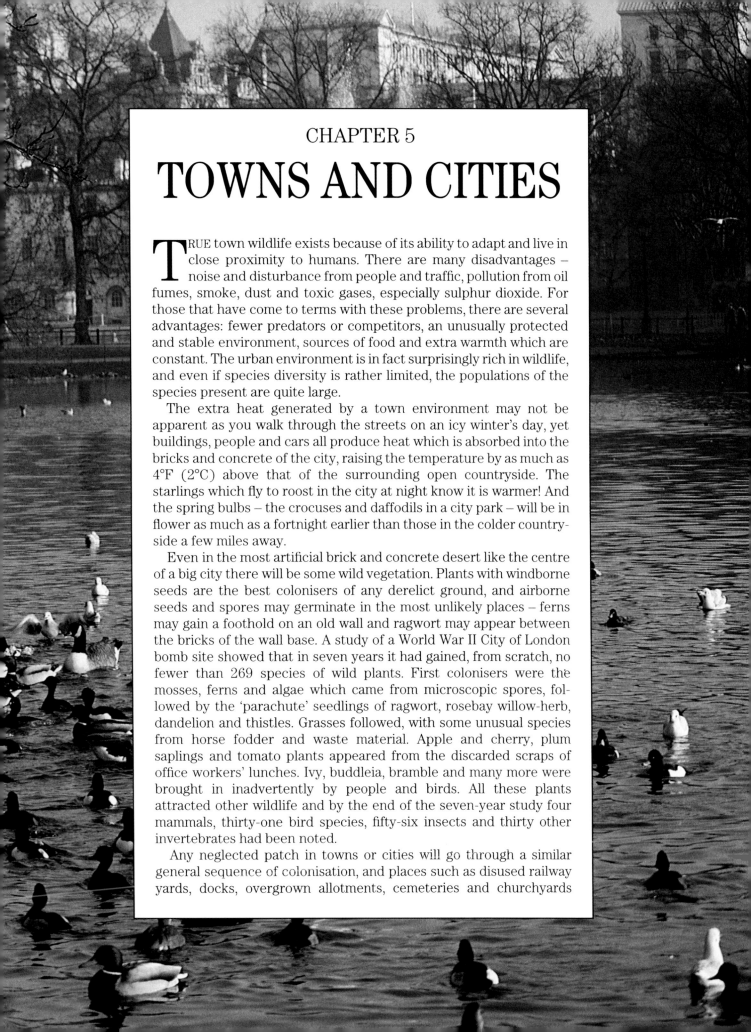

CHAPTER 5
TOWNS AND CITIES

TRUE town wildlife exists because of its ability to adapt and live in close proximity to humans. There are many disadvantages – noise and disturbance from people and traffic, pollution from oil fumes, smoke, dust and toxic gases, especially sulphur dioxide. For those that have come to terms with these problems, there are several advantages: fewer predators or competitors, an unusually protected and stable environment, sources of food and extra warmth which are constant. The urban environment is in fact surprisingly rich in wildlife, and even if species diversity is rather limited, the populations of the species present are quite large.

The extra heat generated by a town environment may not be apparent as you walk through the streets on an icy winter's day, yet buildings, people and cars all produce heat which is absorbed into the bricks and concrete of the city, raising the temperature by as much as 4°F (2°C) above that of the surrounding open countryside. The starlings which fly to roost in the city at night know it is warmer! And the spring bulbs – the crocuses and daffodils in a city park – will be in flower as much as a fortnight earlier than those in the colder countryside a few miles away.

Even in the most artificial brick and concrete desert like the centre of a big city there will be some wild vegetation. Plants with windborne seeds are the best colonisers of any derelict ground, and airborne seeds and spores may germinate in the most unlikely places – ferns may gain a foothold on an old wall and ragwort may appear between the bricks of the wall base. A study of a World War II City of London bomb site showed that in seven years it had gained, from scratch, no fewer than 269 species of wild plants. First colonisers were the mosses, ferns and algae which came from microscopic spores, followed by the 'parachute' seedlings of ragwort, rosebay willow-herb, dandelion and thistles. Grasses followed, with some unusual species from horse fodder and waste material. Apple and cherry, plum saplings and tomato plants appeared from the discarded scraps of office workers' lunches. Ivy, buddleia, bramble and many more were brought in inadvertently by people and birds. All these plants attracted other wildlife and by the end of the seven-year study four mammals, thirty-one bird species, fifty-six insects and thirty other invertebrates had been noted.

Any neglected patch in towns or cities will go through a similar general sequence of colonisation, and places such as disused railway yards, docks, overgrown allotments, cemeteries and churchyards

PREVIOUS PAGE
St James's Park, London, in winter, when wild duck and gulls join the resident birds in an effort to find food. The practice of ringing has shown that some of the gulls have come from as far afield as Russia.

LEFT
A blackbird perches on a gravestone. Quiet old churchyards make good town and city wildlife refuges for a wide variety of wildlife. However, cats always present a grave danger to blackbirds, while dogs and humans are a constant disturbance to other species.

become havens for wildlife. Some of these, especially those with extra environmental features such as pools and marshy ground, are among the richest of all urban habitats and some have even been recognised by the Nature Conservancy Council as Sites of Special Scientific Interest (SSSIs).

While many of these wild, overgrown patches of land exist in our cities, if they are isolated, it may be quite difficult for wildlife to get to them across the large, intervening stretches of urban desert. However, our own highways are also used by animals and birds: railway embankments, canals, rivers and hedged roadside verges all provide access right into the hearts of our towns and cities. Railway embankments in particular provide an undisturbed habitat of grass, shrubs and flowers, and harbour insects, birds, small mammals and even larger species, especially foxes. Furthermore many of our towns and cities are built on rivers, and many creatures will find their way along a river or a canal right into the built-up areas, especially birds and wetland mammals, like the water voles on the Birmingham Canal or in Salisbury town centre.

Birds are especially attracted to water, and gulls, ducks, grebes, swans and even herons and kingfishers use city rivers throughout the year. Gulls are now a notable winter feature of towns and cities; they have taken to moving up river from the coast, finding both shelter and food in the warmer cityscape. Large winter roosts of gulls, involving all the common species but especially the black-headed gull, occur on reservoirs now built on the outskirts of many large urban areas. Some of these roosts may run to 50,000 birds, and it is difficult to realise that less than a hundred years ago the gull was virtually unknown in towns.

To migrant birds flying overhead, the reservoirs, sewage farms and gravel pits must appear as oases in the urban landscape. These regularly attract a wide variety of birds, including waders like ruff and greenshank, ducks like migrant garganey and teal, passing flocks of wild geese, swans and terns, and wagtails, warblers and flycatchers which feed on the multitude of insects. In winter the same artificial wetlands provide undisturbed sites for large flocks of ducks, coots, grebes, geese, swans and gulls – undisturbed, that is, provided they have not been exploited for boating, water-skiing or for any other water sports. They are also stocked with fish for anglers, and grebes and cormorants take advantage of this easy food supply.

Refuse tips are often sited on the edge of the town, all too often in an otherwise interesting habitat like an old quarry, marsh or gravel pit. This is cheaper than burning the rubbish but it inevitably destroys the habitat in which it is dumped. Nevertheless, the tips are full of invertebrates and plants which colonise disturbed ground, and also a great deal of edible household scraps which in turn attract scavengers like rats and foxes, starlings, crows and gulls. Large finch flocks may appear in winter after seeds, while in summer, these town-edge tips are good places to watch bats hunting the myriad swarms of insects which emerge from the 'compost'.

Perhaps the most important single habitat is the suburban garden; large beds with a variety of plants and flowers often surround many of the municipal parks and gardens, which together form urban nature reserves of considerable diversity. Old churchyards and overgrown cemeteries are also usually part of the suburban landscape, and birds and animals move freely between these quiet havens and the nearby gardens. Foxes, badgers, squirrels, hedgehogs, mice, voles, tawny owls, blackbirds, thrushes, robins, tits and nuthatches are just a few of the species which live in town gardens all year, joined in summer by swifts and house martins and in winter by redwings, fieldfares and siskins among others.

Pollution has been a major problem in the cities, but is one that is gradually being mastered; it is measurable to some degree by the population movements of certain species. Thus the house martin and the swift are birds that have moved back into Britain's cities, even into the heart of London. Smokeless zones cleaned the air, so that the insect populations – upon which the swifts and martins depend for food – were able to recover. During the smoke-filled Victorian age, peppered moths evolved a melanistic, black colour form for camouflage in the cities; now, the pale country form is slowly edging back toward the city centres.

The problems encountered by wildlife in towns differ, of course, according to the species. For some, like blackbirds, too many cats are the main problem; for ground-nesters such as lapwings, too many dogs and people; for hedge and scrub species, wrens and robins, too much tidiness; and for grassland flowers and bumble bees, too many lawnmowers.

However, efforts at conservation, and public awareness of the various problems, along with a willingness to spend some money on them, are improving in some cities, and more local authorities and individuals are allowing wild areas to flourish in their parks and gardens. A wealth of interesting wildlife lives within our cities, and surveys have shown that over 60 per cent of the people who live there take an interest in it, studying, photographing, watching and feeding the plant and animal species that thrive within its bounds.

Birds of the Suburbs and the City

Britain is unusual among European countries in possessing a huge number of suburban gardens. Despite their artificial nature, these provide abundant food and quiet nesting sites and have become vital in maintaining a reasonable population of a number of bird species in towns. They have, in effect, become a multitude of little nature reserves in their own right and where there are many in the same vicinity, form their own highly individual habitat. Gardens mimic two rural habitats – open woodland, and inland cliffs and crags which are represented by the rows of houses and other buildings. Thus birds which are common in open woodland are highly successful in gardens where the food supply is often more reliable. Gardens have large areas of flowerbeds, kept artificially bare of vegetation for parts of the year. In woodland, ground-feeding birds such as blackbirds, robins and thrushes have to turn over the leaf-litter to get at the bare ground, or rely on the activities of burrowing moles. However, gardeners are an even better substitute for moles, turning the soil and making it easy for birds to find worms and other invertebrates.

The number of bird species which may be seen in the suburbs is large if one includes all the passing migrants. For example in Regent's Park, central London, over 100 species have been recorded while a suburban park in Surrey boasts a list of 110. Among the regular breeding birds, however, the species diversity is quite low, although this is influenced by the proximity of any large tracts of woodland or commons. Some twenty-two common species of woodland and hedgerow birds now regularly breed in the suburbs: wood-pigeon, collared dove, tawny owl, great spotted woodpecker, starling, magpie, jay, rook, wren, dunnock, spotted flycatcher, robin, blackbird, song-thrush, mistle-thrush, coal tit, blue tit, great tit, house sparrow, chaffinch, greenfinch and goldfinch. Others may breed regularly in certain favoured localities; thus

well-wooded gardens are used by sparrowhawks, nuthatches and bullfinches, among others. Seven species of crag- and crevice-nester have also found that conditions in the town are to their liking: the kestrel, jackdaw, carrion crow, pied wagtail, swift, swallow and house martin.

Garden and park birds are helped greatly by the benevolence of the public, who often provide a wide range of wild bird food throughout the winter, as well as many artificial nest sites in the shape of bird boxes. Those gardeners aware of wildlife requirements can also help in several ways: they can provide a nesting site for a robin, out of reach of the local cats simply by leaving the window of the garden shed ajar. And a very tidy garden is less welcoming than one with a thick, straggling honeysuckle or bramble hedge, especially for robins, wrens and dunnocks. Since 1970 the British Trust for Ornithology has conducted the Garden Bird Feeding Survey and new, interested bird gardeners might wish to contribute their observations. The national *Birdwatching* magazine does the same.

Although most of the common garden birds are all-year-round residents, their populations are also greatly increased in winter by the influx of migrants from colder countries; most arriving in late autumn or early winter. If you walk round your garden and local park on the right morning in late October you may discover the whole place to be alive with robins, or very dark-coloured song-thrushes, continental immigrants which have arrived overnight. Robins, for example, currently number 3.5 million pairs in summer, but this rises to about 10 million in winter with chicks and immigrants.

In late autumn the high-pitched flight calls of thrushes and redwings may be heard as they pass

A song thrush at its nest in a shrub on a suburban house wall. It forages in the open and in undergrowth for worms, slugs and especially for snails which it smashes on chosen 'anvil' stones. Thrushes will also feed on the many berried shrubs planted in suburban gardens such as the cotoneaster.

overhead en route to their winter homes in different parts of Britain. These movements continue night and day, and can be seen over many suburban areas in southern, central and eastern Britain. They are heaviest when conditions are favourable: fine clear weather over the Dutch and north German coasts and on the route across England. A fine autumn morning with a light breeze will often produce huge early morning arrivals, with enormous flocks of birds flying west or north west a few hundred feet above the ground. Most numerous are starlings, chaffinches, skylarks, meadow pipits, wood-pigeons and lapwings; on favourable days,

thousands of these birds may pass over one small area. Other foreign visitors may swell the numbers of other species though to a lesser extent: pied and grey wagtails, jackdaws, rooks, linnets, gold- and greenfinches.

The arrivals and departures of summer immigrants may often be seen in suburban gardens and parks, too. In spring, willow warblers, chiff-chaffs and blackcaps may sing briefly from a copse before moving on to more hospitable breeding sites, although sometimes they stay to breed in the more rural districts on the edge of a town. Swallows and sand and house martins will pass in waves; golf

A robin will often choose the most unusual nest sites – like a tin in a toolshed. This cheeky little bird is always willing to make friends with man and will often watch hopefully as he tills the soil. The robin is one of the most common visitors of bird tables in its continual quest for food – cheese is particularly favoured!

courses and playing fields will attract wheatears and whinchats which stop to rest because these more open areas presumably resemble the breeding grounds for which they are heading – the open moorland. The departures in autumn are less noticeable, warblers creeping steadily southwards from garden to garden, bush to bush.

Sometimes siskin numbers reach an obvious peak in late March; these will be migrant birds moving through the suburban gardens en route to their northern breeding areas, although the siskin is now breeding well in Britain, too. Normally there are maybe 25,000 pairs, supplemented by a sub-stantial immigration in autumn of continental birds which may raise numbers to as many as 150,000 in winter. An indication of population movements is shown by two birds ringed in a Surrey garden which were retrapped in Lithuania the following summer, and caught again two winters later in the same Surrey garden. One was later found (dead) in Finland.

Urban birds have adapted completely so as to coexist with mankind and his artificial environment. The house sparrow is the best example of this coexistence, and is the most numerous wild bird to be associated with humans – it finds an ex-

Feeding house sparrows in St James's Park, London. Although normally quite wary birds, house sparrows operate in non-territorial 'gangs', cleaning up wherever there are easy pickings on farms, hedgerows, parks, gardens, docks, railways and wasteland of all kinds.

tremely good living in the towns and cities, eating whatever scraps of food appear on the pavements and in the gardens, along with seeds, fruit and invertebrates. Urban sparrows are quite omnivorous. Although sparrows are in fact less numerous in modern cities than they were in the era of horses and spilled corn, they have always preferred habitats near to humans, since the dawn of civilisation.

House sparrows are now primarily town birds. The estimated British breeding population is about 4½ million pairs, and studies have shown a density of 5 birds per 2½ acres (1ha) in towns, compared with 1.4 in rural areas. Some live in places like railway stations and factories, and rarely see the open air at all. Probably the best place to see really tame sparrows is St James's Park in London, where they will happily sit on the hand and eat bread and birdseed. This is unusual behaviour, however, and sparrows elsewhere are very wary birds, far more so than robins or tits which can become exceptionally tame.

The other truly urban bird is the feral pigeon. These are now so much a part of our towns and cities that it is often forgotten that they are descended from the wild rock dove, and were first domesticated some 7,000 years ago. In the early cultures of the Middle East they were regarded as sacred; the Romans and many later peoples considered them with less reverence and used them as a source of food and as messengers; while today, some are used for racing but the vast majority live out their lives in the city streets.

Wild rock doves live on remote cliffs, breeding in cracks and crevices. The urban pigeon must, therefore, see the cityscape as a series of artificial cliffs and canyons, using the nooks and crannies in buildings, bridges, railway stations and churches as nesting sites. Urban pigeons feed on bread, seeds, vegetable matter from markets and any scraps left in the city streets. They roost together on roofs for warmth, and where they congregate is usually an indication as to which building is best heated! Because they enjoy a year-round food supply they breed all year round, too.

Trafalgar Square in London is famous for its pigeons, and feeding them is a favourite pastime for tourists and Londoners alike; yet their presence in such numbers is causing considerable problems in some places – for example, during a recent clean-up of government buildings over fifty tons of pigeon droppings were removed from the Whitehall roofs. Like sparrows, pigeons have adapted totally to life with people, sometimes even hopping on to a tube train at one station and hopping out at the next!

Starlings are very common town birds. In the open countryside they nest in tree holes; in the city they use holes in buildings. They are most frequent where there are areas of grass nearby such as parks, playing fields and lawns which they can probe for invertebrate food. They will also come and take bread from bird tables. In winter, huge numbers will roost on the warm ledges of city buildings; they solve their feeding problem by commuting out to the suburbs and beyond every day – London starlings are known to commute 15–20 miles (24–32km) each way from the city centre. Autumn roosts often start in trees but move to buildings, which are warmer, with the onset of cold weather and leaf fall. The highest breeding populations are found in suburban areas, then in woodland; it has been estimated that the starling occupies 97 per cent of the whole of this country, with perhaps three to four million pairs. This figure is greatly increased by continental arrivals in late autumn and winter when the population has been estimated at a staggering 35–40 million birds.

Some of the birds of scrub and woodland have adapted very happily to living in city centres: wood pigeons, for example, prefer to build their nests high up in city trees, and even use the plane trees in the very heart of London. In the country the wood pigeon is extremely wary of man because it is so often shot at, but city birds have lost this fear and can be approached closely.

Blackbirds are more numerous in city centres than song-thrushes because they have proved more adaptable. They have discovered how to push their nests into many a man-made space, and have changed their diet to include a variety of scraps as well as the more usual invertebrates and berries. The song-thrush has not been able to cope with the changes needed and is much rarer in the city, only increasing in numbers further out in the larger suburban gardens.

Wrens and robins breed in ivy-clad walls, while blue tits will use a variety of man-made cavities as

House martins now breed further into town and inner city environments because, with tighter controls on smoke pollution, their insect food has returned to built-up areas. Originally cliff-nesters, they have adapted perfectly to living among buildings, constructing cup-shaped nests of mud lined with straw and feathers under the eaves of houses.

well as the more usual tree holes. Great tits are less numerous and seem to need real tree holes, although nest boxes in city parks and gardens can help considerably.

If the town centre boasts a park lake or some other waterway, there will almost certainly be some mallard, which have adapted to nesting in the most unlikely situations. For example, one pair nests on the roof of the Admiralty in Whitehall, London – when it is time to leave, the ducklings jump to the ground just as they would from a riverside tree, and are then escorted by their mother and a policeman across the busy road and into St James's Park!

In summer there will almost certainly be swifts and house martins swooping about in the skies above town or city, summer visitors from Africa. They are now seen far more frequently above the cities than they were in the past, as the air is cleaner and there are enough flying insects to keep them supplied with food. Swifts nest in the many cavities occurring in house roofs, while house martins plaster their mud nests under the eaves. In London, the most urban of the house martins nest under the eaves of Belgravia and collect mud for their nests from pools by the Serpentine in Hyde Park. A few predatory birds may breed in the city. Kestrels use the 'cliff faces' provided by tower blocks, old buildings and church towers in which to nest, and have also changed their diet; in the countryside they live on small mammals and beetles, whereas in town they prey on sparrows. Most large towns will have no more than one or two pairs; central London supports about five pairs. Where there are large trees, there may be tawny owls, feeding largely on sparrows and starlings caught at their night-time roosts.

The past forty years have seen a dramatic rise in

gull numbers, and many seaside towns now have colonies of herring gulls among the chimneypots. It has been shown that gulls in this artificial environment produce more fledged chicks than on their natural cliff home, because fledglings on the rooftops are less likely to fall prey to neighbouring adult birds, should they stray into their territory.

Magpies have increased markedly in towns in the past thirty-five years; in some northern cities they have learned that the milkman leaves cartons of fresh eggs on doorsteps, and that a magpie's beak makes short work of a cardboard carton. They also breed more successfully in towns where the climate is warmer, enabling them to nest a week or two earlier than those in the open countryside where temperatures are lower, and to continue later into autumn, fitting in an extra clutch if the first clutch is lost.

Wildflowers in Towns and Cities

Wild plants thrive in the smallest nooks and crannies of the city and in wide variety, including many exotic species that man has unwittingly introduced. There are also quiet, undisturbed corners that harbour many plants – and animals, too – like railway embankments, isolated derelict sites, quiet cemeteries and overgrown churchyards. City plants fall into two groups: those that colonise and grow quickly in disturbed ground; and more permanent plants in stable locations.

The most notable opportunist must be the rosebay willow-herb, first recorded in Britain a century ago as an imported garden plant from America. It was one of the first species to cover the London bomb sites with flower after World War II, and readily colonises bare, fire-scorched ground – hence its other name of fireweed. Its success is due to its seeds which are almost weightless, borne on fluffy 'parachutes' and able to travel great distances, sometimes several hundred miles. A single plant may produce 80,000 seeds which all effectively disperse, though studies show that only one per cent will travel more than 6 miles (10km).

Oxford ragwort is a similar success story, and has become a familiar plant of waste ground, railway banks and old walls. It is a native of volcanic areas of Sicily, but escaped from the Oxford Botanic Gardens at the end of the eighteenth century. It produces thousands of tiny airborne seeds in the same manner as rosebay. These have been blown or carried along every single railway track and roadside fashioned by man over the past 200 years, and Oxford ragwort is now a common plant throughout Britain and northern Europe. The lime-filled mortar of old walls, sidings and rubble sites are somewhat similar to its original home on the volcanic ashes.

Derelict urban sites often produce a splendid show of colourful wildflowers throughout the season. Many of the fast colonisers like rosebay and ragwort arrive on windborne seeds. Once down and germinated, many of these plants can also spread by vegetative reproduction, like the underground rhizomes of coltsfoot. Common wasteland plants include dandelion, hawk's-beard, groundsel, sow thistles and other thistles, as well as coltsfoot, rosebay and other willow-herbs and ragwort.

The plant colonisers of waste ground are a very mixed bunch – natives and aliens, garden escapees, arable-land weeds, and plants which would normally be found on landslips and scree slopes. Several species produce dense, tall growth, like stinging nettles, fat hen, common orache and various docks. Two common alien species are thorn apple and Japanese knotweed. The thorn apple is highly poisonous, with huge white trumpet-like flowers, while knotweed is a voracious coloniser and a thoroughly regrettable introduction. Fat hen displays a further adaptation towards survival on disturbed ground by producing abundant seeds which can remain dormant for varying lengths of time, a policy which ensures that some survive to

With the help of its tiny 'parachute' seeds, Oxford ragwort has spread throughout the country in the last two hundred years. The ragwort has been particularly successful along railway lines and embankments, colonising the clinker ash between the rails which so closely resembles the volcanic ash of its origins in Sicily.

Buddleia attracts many insects, particularly butterflies and is thus known to many as the 'butterfly bush'. On warm sunny days these colourful shrubs can be a moving mass of red admiral, peacock, tortoiseshell and painted lady butterflies.

germinate when conditions are suitable.

Several waste-ground plants have seeds with hooks or barbs so that, like burrs, they stick to birds and mammals – including humans – which carry them until they brush off against the vegetation of, hopefully, a new site. Goosegrass, wood burdock, lesser burdock and enchanter's nightshade are all very common plants of this type.

Plants growing in places where they are likely to be stepped upon survive best if they are tough, leathery and low-growing. It often helps if their leaves are arranged in a flattened rosette, like daisies and plantains which both flourish in areas of mown grass – lawns and playing fields – or on paths or between stones. Pineappleweed (rayless mayweed) is just such a one, a small, daisy-like plant which smells strongly of pineapples when crushed. It is now widespread along roadsides and tracks, its seeds having been carried and dispersed by tyres and boots since it first appeared at the turn of the century when it was imported accidentally from north-east Asia.

Old, battered walls are occupied in towns by many plants which need dry, well-drained, mountainside conditions, especially those with lime mortar between the bricks. Some are native, like stonecrop, wall pennywort and wall rocket, while others were introduced either accidentally or as garden plants, like wallflower, ivy-leaved toadflax, yellow corydalis and red valerian. Many climbing plants grow on town walls, but ivy and bindweeds are the most frequent. Alien species such as the quick-growing Russian vine may also be found.

Once an area of wasteland has settled down, the first colonisers are shaded out by larger and more permanent plants like bramble, silver birch and buddleia. Only introduced from the Far East at the end of the last century, buddleia can establish itself even on old walls, its roots able to take hold in the nutrient-filled cracks in the stonework.

Old canals and pools on industrial wasteland are often rich in marshland plants, some growing on damp brickwork, like skullcap and gipsywort, while others – like yellow flag, meadowsweet and water dock – grow in or alongside the water. Great hogweed and Himalayan balsam might well be present, too, both of them introductions which are unfortunate, as they are so invasive that native plants are often driven out.

Tall grasses are rare in towns, and usually occur only along the edges of the quietest, most undisturbed sites in places such as cemeteries, railway banks and old churchyards. These act like nature reserves in towns and cities, allowing birds and mammals alike a solitude difficult to match even in the countryside. In so many cases neglected, these railway banks and old churchyards often become overgrown with shrubs like elder, hawthorn, bramble and briar, but then provide food and cover for thrushes, blackbirds, magpies, wood-pigeons, rabbits, foxes and even badgers. The long grass shelters both shrews and field voles, while in high summer ox-eye daisies, knapweed, cow parsley, golden rod, red valerian and buddleia are just a few of the pollen-filled flowers which bring insects, especially butterflies, from far and wide.

On the weather-worn gravestones of city graveyards grow patches of lichens, though so many of these are susceptible to atmospheric pollution. Xanthoria is a bright orange leafy-lobed lichen, while *Leucanora* sp are usually grey-green. There are hundreds of species and their presence or absence indicates the level of pollution in the area.

Mammals in Towns and Cities

Even eight thousand years ago, when early man was only just starting to build permanent settlements, he shared his dwelling-places with a wild mammal: house mice are recorded as infesting buildings in ancient Eygpt and in Turkey. Some mammals have become so attuned to human society that now they are found most often in the cities and towns, and in the case of the smaller species, populations are highest in the houses themselves. Otherwise, a few species tolerate humans but only at a respectful distance and will accept their presence provided they show no ill-will; thus the badger may live in harmony with a suburban human community, falling foul only of the trap and speeding car, and illegal badger-diggers. With a little ingenuity, most of these city mammals are easy to see; although the mammal which has most direct effect on the rest of the town's wildlife is the household cat. It has been estimated that cats kill 100 million wild birds and mammals every year in Britain alone.

Brown rats and house mice are by far the most numerous of urban mammals, although recent surveys revealed that the number of city-dwelling brown rats was quite low; however, 1988 saw a dramatic rise in the number of households requesting local authority rat-removal services, so perhaps the trend is reversing. Large colonies occur in sewers and on rubbish tips, but surveys show that only some 3 per cent of urban premises are infested with rats and that 80 per cent of these are outside the buildings. Urban rats in fact tend to live in small scattered groups and are most likely to be seen nibbling food put out for birds on garden lawns.

The brown rat only arrived in Britain in the eighteenth century but soon almost completely supplanted the longer-tailed, larger-eared black rat which is now confined to port and dock areas; and even in these it is nearly extinct. Rats have an extremely wide diet and the brown rat has adapted very well to living in a variety of different habitats,

The hedgehog is a helpful visitor to suburban gardens, ridding them of pests like slugs and snails. Many hedgehogs are killed on the roads each year, but many more die in gardens – especially in slippery sided, plastic-lined ponds, and also in bonfires where piles of dead leaves may have attracted a hibernating animal.

BELOW
The brown rat only arrived in Britain in the eighteenth century, but has now become widespread throughout the country. It is a prolific breeder and is generally regarded as one of the few really undesirable wild animals because of the economic damage and disease it causes.

these two factors ensuring its success. The brown rat's weight is variable but can be over 1lb (500g), so making it on average two and a half times the weight of the black rat, which averages just under ½lb (200g). The brown rat eats the equivalent of 1–1½oz (30–40g) of wheat daily but will tackle anything including many house mice in corn ricks. The black rat confines its living habits more to buildings, thus making it easier to eradicate. Both species do great damage to stored foods, and the black rat carries the flea *Xenopsylla choepis*, the main transmitter of plague and typhus. Five litters of babies a year are normal, and a rat is mature enough to breed from three months onwards; furthermore, if food is available, breeding will continue throughout the year. The brown rat survives best in temperate climates, while the black rat thrives best in the tropics.

The house mouse is another versatile species which is found all over the world and in all sorts of places: on mountains, down mines, in heated warehouses and in commercial cool-houses, as well as being firmly established in human dwellings – as we have seen, it has been a commensal of man for at least 8,000 years. Its average weight is just under 1oz (28g) for females, and ¾oz (22g)

for males, but this varies widely. It has only a small foraging range, of about 50sq ft (4.6sq m). The female produces up to ten litters a year with an average of 5.6 young in each litter; these are weaned at eighteen days and are able to breed themselves in a month. In a cool-house the house mouse will grow a thicker coat and still produce an average of six litters a year. House mice like to live close to their food supply, in garages, storage cupboards and under kitchen floorboards, and in a house may have well established runs all round the edge of each room.

All these urban rodents are preyed upon in turn by cats, dogs, foxes and tawny owls. Cats also prey upon squirrels, although the squirrel has a savage bite and many cats learn to be rather more wary when hunting them than they would smaller rodents. The red squirrel is now only likely to be seen in isolated gardens in Scotland and the north west; the grey squirrel, however, is the most frequently seen urban mammal of all, and makes a fat living from food put out by humans, usually for the birds. Despite its attractive antics in climbing after peanut feeders, the grey squirrel can do a great deal of damage, eating birds' eggs and chicks, fruit and vegetable crops and tree buds and bark. It will strip off tree bark in spring to get at the sweet, vitamin-filled sappy layers beneath and can easily kill a tree by 'ring-barking' it (chewing off a complete ring of bark around a tree trunk), thereby cutting off the supply of sap which rises up the bark to feed the upper tree.

After the diurnal squirrel, the next most familiar garden mammal is the hedgehog, most probably seen in the dusk, especially if it has learned to come for a bowl of bread and milk; it will also readily eat dinner scraps and meaty tinned pet foods. Research has shown that hedgehogs simply treat this food as a supplement to their more normal diet of beetles, leatherjackets, caterpillars, slugs and other destructive garden pests. For gardeners, regular visits by the neighbourhood hedgehogs are worth encouraging – but they can be poisoned by slug pellets. They may travel some distance in their search for food, and very often have a regular night-time route, popping in and out of garden gates as they make a nightly survey of 'their' patch – so your one faithful hedgehog may in fact be several passing through. Marking with a little blob of

paint on the spines has shown that as many as eight or ten may move in and out of a favourite garden.

Hedgehogs hibernate from November round to early April and need to lay down a lot of fat in autumn as an energy store to survive the winter; late-born hedgehogs which are under 1lb (0.45kg) in weight by autumn are unlikely to survive. If very small hedgehogs are found after about mid-September they can be kept very successfully indoors, alive, awake and well fed on dog or cat food and scraps until the spring; the hedgehog does not have to hibernate provided it is kept well fed and warm (above about 50°F/10°C). The normal weight of the hedgehog varies widely according to its age and the season, but the average autumn adult is about 2½lb (1,200g). There is usually one litter per year, although there are two peaks of pregnancy, in May/June and Sept, with four to five young which do not mature until the following year.

The expansion of the fox population into the suburbs is a fairly recent and fascinating zoological phenomenon; it is paralleled by the coyote in some parts of North America, a similarly adaptable mammal. Foxes began to colonise London, primarily in its outer suburbs, just before the start of World War II. It was once thought that this invasion took place after myxomatosis decimated the rabbit in the rural areas in 1953 but it seems much more likely that this was only an additional factor in their colonisation of the cities which was already under way. In London the main reason seems to have been the expansion of the city itself, and the way this happened.

In the late 1930s vast areas around Britain's larger cities were developed into suburban housing estates, full of nice quiet gardens; and the same sort of thing was happening in smaller towns, too. It was this sort of environment that the fox learned to exploit and nowadays it is probably easier to see a wild fox in the suburbs than it is in the countryside. They may not be quite so numerous right in the heart of the big cities, but in London foxes still live and produce cubs in Brixton, Peckham, Wandsworth and Islington. The fox first came to enter the city down the quiet 'corridors' provided by the railways and many foxes still use undisturbed railway embankments for their earths, although now they will raise their litters in a wide variety of locations. The favourite site is probably beneath a garden shed at the bottom of a typical suburban patch.

Most foxes emerge around dusk and then 'do the rounds' of the places providing their prime food supply. They soon get to know where food is regularly available and people who have been feeding their local foxes each day may see the whole family waiting patiently on the lawn for food to appear. During the daytime foxes are usually to be found asleep, in a quiet and undisturbed corner, and often in the sunshine, if there is any. Radio tracking has shown that they are active at intervals through the night until dawn.

It is a myth that city foxes are all half-starved and mange-ridden – indeed they are as sleek and well fed as their country cousins. It is the vixens with cubs that are often thin and very scruffy, but this is most likely to be due to their losing fur in their summer moults. Radio tracking has also shown that some rural foxes actually commute to feed in the suburbs at night, returning at dawn to the quieter countryside. Out in the countryside a fox may have a home range of as much as 2,500 acres (1,000ha) but when plenty of food is available in towns this may be reduced to only 75–100 acres (30–40ha).

Many people worry that their pets may be attacked by the foxes in the area. And certainly, foxes are opportunist feeders; if a free meal in the shape of a backyard hen appears it is likely to be accepted! Cats on the other hand are hardly ever in danger, and only kittens run any real risk. The occasional cat found at a fox earth is usually one which has been run over and collected from the

Fox cubs feeding from food put out in a suburban garden, where they have a good, easy life. Near humans there are always scraps to scavenge and dustbins to investigate, and in return they do man the favour of keeping the rat and mouse population under control.

gutter. Extensive studies, by stomach analysis, of the food of urban foxes show that a wide variety of items is taken as food: worms, 12 per cent; wild mammals, 13 per cent; wild birds, 14 per cent; scavenged meat bones, 24 per cent, and other scavenged items, 11 per cent are the main food sources. This analysis also included 2.9 per cent household and domestic animals, mainly chickens, pigeons and park ducks; but no cats.

The fox mating season is in the depths of winter when they are often noisy neighbours, the screams and the loud triple bark being well known. Both dog and vixen make these noises despite country views to the contrary. The cubs are born from early spring onwards and are above ground by April. Their boisterous play in quiet gardens gives entertainment to many people, and brings a touch of the wild to their city lives.

Some suburban gardens are lucky enough to be near a badger sett, and the badgers can often be persuaded to come for food. Badgers are surprisingly common in towns but because they are secretive and nocturnal, their presence, despite their size, goes unnoticed by all except the observant few. However, provide a wide variety of food and the local badgers will become addicted to their nightly helping of peanuts, sultanas and honey sandwiches – and much more. A badger will eat far more food than a blue tit, so feeding them peanuts can become expensive! Try a proprietary dried vegetarian dog food, which is probably a better diet, although water must be made available too. Holes in the lawn may mean that the badgers have been eating worms or leatherjackets, or even daffodil bulbs!

Just as secretive as the badger, but now found in gardens, parks and woodland over much of southern and central England, is a tiny deer: the muntjac. Its small, curved brown shape slipping through the shrubbery is not often seen and is easily mistaken for a cat or a fox, yet it is certainly present in a great number of suburban areas. It can, unfortunately, be destructive in gardens, eating shrubs and suchlike. Little is known about its habits in the suburbs, but it has proved most adaptable in habitat, diet and reproduction, it seems to breed at any time throughout the year, and adaptability is the key factor in the success of all urban mammals.

Insects and other Invertebrates in Towns and Cities

Many of the thousands of insect and invertebrate species in Britain find the town or city a hospitable habitat. Walk in the garden on a bright summer's day and the wealth of insect life will be only too apparent; a stroll in the evening will reveal the nighttime insect life, the host of invertebrates which take over from the diurnal species – the woodlice, worms, snails, slugs and especially the moths.

Some species of invertebrate, like cockroaches, are found almost entirely within the bounds of man-made buildings, although most that occur in towns also occur in rural areas. In the towns, numbers may be localised and very large, purely because of the artificial nature of the habitat; for example, there may be hundreds of the large and colourful caterpillar of the privet hawk-moth on one garden privet hedge, simply because this is the only available privet – in the open countryside the eggs would be better dispersed because there would probably be more privet bushes scattered over a wider area.

In parks, gardens and wasteland the butterfly is the most noticeable insect, although species diversity is limited in towns as many larval food plants are not available. The most frequently noted species are the large and small whites whose larvae feed on cabbages on allotments and in gardens; small tortoiseshells and peacocks, with larvae on isolated clumps of stinging nettles; holly blues which like quiet parks and gardens with ivy-clad walls and holly bushes; gatekeepers and meadow

A peacock butterfly on buddleia. A peacock may live for as long as ten months as it hibernates during the winter. The stinging nettle is its only natural foodplant and one often sees masses of the butterfly's larvae deposited on its leaves.

browns on wastelands covered in long grasses; and speckled woods and brimstones in lightly wooded suburbs. Many of these species are attracted to garden blossom for nectar, especially flowers like buddleia, ice-plant, Michaelmas daisy and ivy blossom. In the years when the red admiral migrates from the Mediterranean in good numbers, they too appear on flowers in gardens and often feed on rotting windfall fruit in autumn.

When an insect's food plant invades an area in quantity, then the insect itself will not be far behind; in the town this is particularly true of the different species of hawk-moth. For example the willow-herbs, and especially the rosebay, are the food plants of the elephant hawk-moth, a spectacular midsummer insect whose caterpillars reach 3in (8cm) in length. Since it provides such a juicy morsel for birds, the caterpillar has a dramatic eye-spot coloration as a defensive measure. In a similar way, another of the large hawk-moths, the eyed hawk, also has eye spots on its lower wings; the eyed hawk larvae feed on sallows and willows. Two other widespread town hawk-moths are the lime and the poplar hawk, the caterpillars

eating large holes in the new spring leaves of their respective trees, providing an easy way to spot them against the bright sky. Privet hawk caterpillars betray their presence because their black droppings are so noticeable on the pavement below the privet hedge.

Oxford ragwort is home to the orange and black caterpillars of the bright coloured red and black cinnabar moth. The ragwort produces toxins which accumulate in the body of the caterpillar; the caterpillar has evolved a warning colour pattern to advertise this, and this protects it against attack from birds.

Other town moths include the puss-moth, whose larvae feed on sallows and willows, and are a bizarre shape, each one possessing a 'tail' with which it can whip attacking parasitic flies. There is also the garden tiger-moth, with its familiar 'woolly-bear' caterpillars which feed on many garden plants; and the house clothes moth which attacks woollen clothing.

The town moth which has been most informative in its development is the peppered moth, a remarkable example of evolution in action. Its nor-

mal colour is pale and lightly speckled, but with the expansion of the industrial areas of the Midlands in the early nineteenth century a black form began to appear, and by 1900 the black peppered moth was predominant in industrial regions. Furthermore experiments in the 1950s showed that in soot-blackened Birmingham 90 per cent of the moths were black, while in unpolluted Dorset 95 per cent were pale. Even more interesting is the fact that now that smoke-control legislation produces cleaner buildings and trees, the black version of the moth is becoming less frequent. Birds feed on these moths, and careful observation has revealed that each colour form does survive much better in its own environment since it is more effectively camouflaged, the black one in the sooty areas, the pale one in the 'clean' countryside. Several other species now show industrial melanism including the pale brindled beauty, the scalloped hazel and the green-brindled crescent moths; two-spot ladybirds, zebra and wolf spiders also show this genetic change.

Garden insects are well known and to the gardener, fall into two main categories – pests and others. Insecticide sprays, of course, are indiscriminate and kill both types. Aphids such as green- and blackfly are abundant; they carry many plant diseases and attack many plants, but especially roses – they feed by sinking their mouth parts into the plant stem and sucking out the sugary sap. Any undigested residue is excreted as sweet 'honeydew' droplets, and these are used, even 'farmed', by many ants. The aphid population is controlled naturally by winter temperatures and many die, but the eggs produced in autumn by winged females are more resistant and survive. Throughout the summer, many generations of aphids are born live from wingless females which do not require male fertilisation, a process known as parthenogenesis. Only in autumn are males produced, to fertilise the overwintering eggs. Aphids are themselves preyed upon by ladybirds and their larvae, and by lacewings and hoverfly larvae, all of which are common garden insects, though obviously the population of the predatory species depends upon the population of its food supply. Aphids also provide food for many insectivorous birds like blue tits and warblers.

Frog-hoppers are bugs whose larvae envelope

themselves in what is commonly known as 'cuckoo-spit', a frothy substance which protects them against predators, and also prevents them from drying out. They belong to the same order of insects as the aphids (the Homoptera) and feed in the same way. There are some 350 species in Britain.

In high summer the most numerous bee to be seen about town is the honey bee, notable as one of the few insects to have been domesticated. Each colony contains some 50,000 bees; most of these are workers, which are sterile females, but there will be a few males, called drones, and one large breeding female, the queen. The workers collect nectar, and produce and store the honey which is needed to feed the larvae and tide the colony through the winter. The queen mates once with one of the drones, which are produced in very small numbers for this sole purpose. The queen may live for three to four years, and lays eggs throughout her life; the worker bees, however,

Ladybirds feed on aphids, and are therefore of great assistance to town gardeners. Great care should be taken not to destroy them when spraying plants with insecticide. Members of the ladybird family of beetles are usually red or yellow with black spots. Their markings are, however, very variable, ranging from two to twenty-two spots.

only survive for a few months. In a good season a single hive may produce 440lb (200kg) of honey. As the bees need a continuous supply of flowers throughout the summer, and so do proud gardeners, the bees often benefit and do very well in towns.

Only a handful of wasp species are social insects in the same way as bees, and these form colonies in 'paper' nests made of chewed wood pulp. For most of the summer wasps are beneficial because they take many of the caterpillars considered as pests as food for their larvae. However, in late summer the queen wasp stops producing grubs, and the sweet secretions which feed the workers also stop; it is then that the wasp becomes a plague on its human neighbours, descending on jams and sweetmeats, trying to satisfy its need for a sweet food.

The hoverfly is also seen frequently in summer, prominent amongst the town insects attracted to garden flowers. There are some 250 British species, all with remarkable powers of flight which enable them to hover for several minutes in front of flowers while they search for nectar. Many mimic various species of bee and wasp, and thus gain a measure of protection against attack by birds.

During the night the invertebrate life which emerges is quite different in type and character: night-flying moths become the pollinators, and it is the turn of snails and slugs, earwigs and woodlice to become active, together with spiders and nocturnal beetles such as maybugs. Some of these night-time creatures have been introduced, such as the large American cockroach and the house cricket from North Africa which are both found in many buildings. A more unusual foreigner is the European scorpion, which has established small colonies in the docks and even in the London Underground; it is the size of a 5p piece, and thankfully is relatively harmless!

CHAPTER 6
FRESHWATER HABITATS

RIVERS, streams, lakes, ponds, fens and bogs all support a rich and varied wildlife; where it is at its richest it is more diverse and exists in greater quantity than on any of the dry land habitats. On the other hand, some freshwater lakes are so sterile that they resemble deserts; compare the reed-fringed, lily-filled lake of the southern counties with the clear, cold waters of a northern Scottish loch. This is because plants, in particular, need nutrients – nitrogen, phosphorus and potassium – and water plants need these dissolved in the water; in the cool mountains of the north and west where the water runs over hard, ancient rocks, not many nutrients are dissolved, therefore there are only a few plants and even fewer animals. The water that runs over the softer limey rocks of lowland Britain dissolves great quantities of these essential plant foods and hence the rivers and ponds are filled with life – a simple reflection of the geology of the base rocks.

The eco-systems of the freshwater habitat are as complex and varied as those on the land. Furthermore, the distinction and differences involved between still and moving waters are clear and all-important – between ponds, lakes and canals on the one hand, and rivers and streams on the other. The freshwater habitat is also complex because the animals and plants in it are not solely dependent upon the water. For example, some animals are emergent and breathe air, and some plants must have their roots in water while the rest of the plant needs to be raised above it. Most plants raise their flowers, at least, above the water surface so they may be pollinated by flying insects.

Defining the various freshwater habitats, for example identifying a pond as opposed to a lake, is not simple and has little scientific basis. A pond is basically a small body of still, fresh water, in which there is little difference between the surface temperature and that of the bottom. A lake is a larger body of water, too deep for emergent plants to grow in the middle and with marked differences between the temperature of the surface and the bottom. Most so-called suburban park 'lakes' are really large ponds.

No two ponds are alike, and every pond has a different collection of plants and animals. The most important single factor – after the nutrient supply – which dominates the life of a pond is the singular property possessed by water of being most dense at 40°F (4°C). Once the surface layers cool below 40°F (4°C) and freeze, they float on the denser water beneath and so insulate it. The bottom therefore remains at 40°F (4°C), providing a haven for creatures which could not survive in ice. If water were most dense at 32°F (0°C) the pond would freeze solid in winter.

In deep lakes this temperature gradient is noticeable throughout the year. There is a shallow upper layer known as the 'epilimnion' which floats on the cold, deeper layer called the 'hypolimnion'; there is a marked temperature change at the junction of the two layers called a 'thermocline'. Basically, the warm 'summer' water of the upper layer floats on the dense lower layer, a stratification which may remain unchanged until late autumn when the upper layer cools – this may cause nutrients to be trapped for long periods in the lower layer, and the lower layer may become anoxic (lacking in oxygen). Because light only penetrates the upper layer to a dozen feet or so at most (a few metres) it is here that all the algal, or plant growth takes place. Although a pond may be full of visible plants the algae are the most important food source for the majority of the pond animals.

Lakes in areas where the rock is old and hard, which lack nutrients and have little in the way of plant or animal life, are called 'oligotrophic' lakes, (literally 'little food'); those rich in nutrients and therefore full of life are called 'eutrophic' lakes. Ironically, many lakes and ponds now suffer badly from enrichment caused by the run-off of nitrogen-rich farmland fertilisers. These produce large-scale blooms of algae which, while providing food and oxygen for lake animals during daytime, actually remove an even larger amount of oxygen at night. Therefore lakes which are enriched in this way may eventually 'die' from lack of oxygen, leading to the wholesale death of fish, pond animals and plants.

On the other hand, man may sometimes improve the environment – albeit unintentionally, as in the case of gravel and clay extraction for building, for example. Initially this may destroy valuable countryside but later, as the pits are worked out and abandoned, they fill with water and eventually provide a thriving wetland habitat. Man-made reservoirs are also a haven for thousands of birds, particularly in winter.

Ponds and lakes have five distinct zones: the pond edge with its swamp plants; the area further out where the emergent, but truly aquatic plants like water-lilies grow; then at a deeper level, the area of submerged plants; the pond bottom; and the pond surface. The latter is interesting, as water molecules are attracted so strongly together that a high degree of surface tension is created, on which many animals actually live. Especially noticeable in this group are pond skaters and whirlygig beetles.

The chain of interdependent wildlife species typical of these stillwater habitats is remarkably rich and varied. Right at the bottom the wide variety of tiny pond plants are consumed by a horde of different species of microscopic animals. These feed the larger and more noticeable creatures such as frogs, newts, insect larvae and fish. These in turn are eaten by a variety of birds like the heron and osprey, and predatory mammals like the otter and the mink, which represent the top of the wetland food chain. The dense growth of emergent plants such as reed, bulrush and yellow flags provides shelter. The plants themselves are not eaten by any of the aquatic animals except the herbivorous water vole; they do, however, conceal many terrestrial insects and their larvae which are preyed upon by several different wetland bird species.

In the southern lowlands eutrophic ponds with a wide margin of reeds and flags will support the reed and sedge warbler, reed bunting and the little grebe, none of which will frequent a bare upland lake or man-made, concrete-banked reservoir. A disused gravel pit with emergent vegetation will also have gained a marshland edge, and therefore act as a suitable breeding ground for all four species in this narrow 'edge' habitat.

In the still waters of ponds and lakes the entire underwater life-cycle depends upon microscopic plants – diatoms and the like – which occur in millions. These cannot exist in the running water of rivers, but are vital to a still-water system. However, the balance in which such a system flourishes is delicate and as we have seen, may be dramatically disturbed by excessive run-off of fertilisers from farmland, resulting in long-term algal blooms which can kill whole aquatic plant communities. The effects of this can be well seen in the Norfolk Broads.

Streams and Rivers

The all-important factor governing the plant and animal life in streams and rivers is the flow and speed of the water. In ponds and lakes the water is still and so there is considerable growth of microscopic plants or phyto-plankton; these in turn support a wide range of animals. In flowing water the phyto-plankton cannot survive in such large quantities and the prime plant food source is detritus, both from bankside plants and, away from the headwaters of the river, from submerged aquatic plants.

The water supplying the rivers comes from rain on the hills, and the familiar V-shaped valleys result from the fast-flowing streams of the uplands which for millions of years have rushed over the hard rock, even though the streams may be reduced to a mere trickle in the valley bottoms in summer.

These hill streams carry away much soil and debris, and so their bed consists of rocks and boulders. Water splashing in a turbulent fashion over such a stream bed becomes filled with oxygen, a characteristic of turbulent water on all streams

PREVIOUS PAGES
Radipole Lake, Weymouth, Dorset. The reed beds of this well-known freshwater nature reserve are densely populated with marshland birds including duck, geese, swans and grebes.

The River Barle, Exmoor, is a fast-flowing hill stream full of trout. The turbulent water is heavily oxygenated so plant and animal life has adapted to suit these conditions: plants are slim and pliable, stone-fly nymphs are found under stones and only powerful swimmers like trout can cope with the current.

and rivers. As the gradient of the stream falls, it begins to drop its load of material, firstly the larger shingle, then the gravels and sand, and finally the silt, on the slow lower reaches of the lowland river.

Where the water flows swiftly, as in the upper reaches of the river, plants and animals must adapt so they are not swept away by the current – in the same way that on land, many species have adapted to avoid being blown away by the wind. The rapid waters of hill streams and torrents do not support very much life. There are no submerged plants, only mosses and liverworts which grow on sheltered boulders in the 'splash zone' of the stream. Animal life is very restricted, usually comprising nothing more than a few stone-fly nymphs which have adapted to their turbulent conditions by clinging to the underside of rocks with their strong legs.

As the current slackens a little the river acquires its first shingle and gravel patches along the rocky bed. The water is still turbulent and full of oxygen and the animal life is still limited, the only surviving species being those that can cope with the high levels of oxygen and the turbulence. Nor are there any submerged plants yet, but the first fish will almost certainly be present: the trout, a strong powerful swimmer designed to cope with the current. Stone-fly nymphs will also occur under the stones. Both these creatures suffer, however, if either summer drought or pollution adversely affects the water flow, because the high levels of oxygen necessary to their survival are then lowered. This hill stage of the river is often called the 'trout beck'. All the lower reaches of the river are also named according to different species of fish most likely to be found in them.

The next stretch of the stream, with the lower gradient and slower water current, though still perhaps in the foothills, is called the 'grayling' or 'minnow zone'. The water is still highly oxygenated, but for the first time the current has slowed enough to allow a few plants to take root. Trout are still abundant and salmonoid fishes predominate, including the salmon itself in some rivers.

Most of the water plants characteristic of these fast-flowing waters are like the water crowfoot, with slim, pliable stems and very finely cut leaves. Instead of attempting to resist the flow of water (which would require massive strength) they have adapted instead so they offer the least resistance to it, trailing out in elongated clumps in the same direction as the current. In summer when the water is at its lowest, they develop broader, floating leaves (so as to utilise all the available sunlight) and then, like most aquatic plants, they raise a mass of white flowers above the water surface for insect pollination. Plants such as pondweeds, with green and insignificant flower spikes, are wind pollinated.

As the river enters the lowlands, there is a fairly narrow, transitional zone where the upland stream really becomes a river and the speed of the current changes markedly. This is known as the 'barbel zone'. Finally, the last and slowest stretch of the lowland river is called the 'bream zone'. Water chemistry and the nature of the underlying rocks becomes more important as the water speed decreases. Most upland streams are oligotrophic, usually flowing over acid rocks; but as the river moves into the lowlands, so it sheds its finer sand and mud enabling plants to take root – and it is the mineral content of the water which plays a significant part in determining what plants may grow. Where the river arises from, or flows over, areas of limestone or chalky rock, the pH value will be high with more nutrients, thereby allowing a greater diversity of species. Also, higher concentrations of calcium in the water support a large number of animals with shells, especially water snails.

The current of the lowland river is slower at the margins, along the banks, than in the middle. Left to itself the river will probably meander down the gently sloping contours in long, sumptuous curves. On bends the current will travel faster on its outer curve, cutting steep banks, whilst on the slower, inner curve it will deposit its load of silt. The river bed is usually of mud and sand, allowing a wide variety of aquatic plants to grow, while along the margins grow emergent plants, with their roots submerged; even the floating aquatic species are

A freshwater reed marsh at Titchwell RSPB reserve, North Norfolk. Coot, moorhen, grebes, bitterns and avocets are just some of the marshland birds to be seen here.

able to survive the very slow water speed. The oxygen levels are much lower, so the animals which occurred in the river's higher reaches are largely absent, although different species of stone-fly may still be found. The fish are 'coarse' species, such as rudd, roach and bream. Detritus is still the main food source as the phyto-plankton are too few to become the main source of food, too small to cope even with a slow current.

Extensive plant growth slows the current still further, causing even the finest silts to be precipitated. However, few rivers are allowed by man to stay in their natural condition. Dredged channels for faster flow, locks, weirs and finally industrial use of river estuaries all change the natural course of a river's life. Perhaps the greatest problem faced by lowland rivers in the 1980s is eutrophication, or enrichment, by excessive fertilisers applied to farmland; clean, unpolluted rivers and streams are of very rare occurrence in England. Rivers in Scotland and Ireland are still clean by comparison.

Marshlands

The plants and animals which occupy the marginal ground between open water and dry land are determined by the degree of wetness. And the character of the marsh or bog is influenced by the climate, the geographical location and the rock type of the area. Succession from open water occurs as follows: in deep water the only plants which occur are fully submerged. However, silt may be gradually deposited as a result of erosion of river banks upstream or because the sides of the open water weather and shelve in. Eventually some of the water will have become shallow enough to allow rooted plants with floating leaves, like water-lilies, to become established. Soon, emergent plants like reeds will be able to grow, further slowing the water currents and allowing still more accumulation of silt, thereby encouraging a build-up of dead plant material.

Slowly the level of the humus-filled silt rises so that it is only flooded in the winter. The reed-swamp plants decrease, and colonising trees such as willow and alder invade the drier ground. From henceforth it is the accumulation of plants and plant remains that is responsible for stabilising further dry ground, and if the ground remains waterlogged these may not decay at all but will form peat instead.

Upland marshes, where the rainfall is high, the soil acid and the rocks hard and impermeable, are true bogs; the main vegetation is sphagnum, or bog-moss, capable of smothering large areas including other vegetation. Sphagnum actively exchanges nutrients for hydrogen ions which are then released into the bog, making it even more acid; a specialised flora comprising only a few species occurs on acid bogs.

Wetlands which form in calcareous or alkaline ground are very different as the water is full of nutrients, and a great variety of plants and animals can thrive there. These rich marshlands are called fens, and originated where reed-swamp took over the open water. Peat accumulates quickly, so the marshy basin, as long as it is left undisturbed, becomes a rich mosaic of fen meadows and wetland trees. Ancient water meadows were used regularly over the centuries for grazing and hay cutting, and are full of unusual fenland plants, many adapted to this ancient farming cycle.

The fenlands of East Anglia once formed the largest area of marsh in Britain, although now only remnants remain. Efficient modern drainage systems have turned the rich, peaty soil into productive farmland, and it is drainage which is the main threat to the all-too-small areas of our remaining freshwater marshlands – although enrichment by farm fertiliser is another major threat. Fenland nature reserves, like those at Wicken and Woodwal-

Woodwalton Fen, Cambridgeshire. Such habitats are home to a wide variety of insects including the pyralid moth, mosquitoes, flies and dragonflies. Fen meadows have been re-created which are full of flowers.

ton in Cambridgeshire, are now several feet (a metre or so) higher than the surrounding dry farmland because the latter, deprived of its water, has shrunk. Elaborate water-pumping regimes are needed to maintain these tiny, precious remnants of the original fens.

Reed-beds provide the most familiar marshland scene, sheltering a great variety of animals, especially birds and insects. However, in 1979/80 the Royal Society for the Protection of Birds ran a survey to assess the situation in England and Wales; Scotland was excluded because very few reed-bed birds breed in its northern climate. Only 109 reed-beds of 5 acres (2ha) or more were found, totalling only 5,683 acres (2,300ha) in the whole country.

Characteristic reed-bed birds include the resident and very rare bittern and the bearded tit; while reed and sedge warblers, and marsh harriers, are summer visitors from Africa. Reed buntings are also numerous in drier areas, and grasshopper warblers, also summer visitors, are found where the drier 'edge' of the reed bed merges into scrub. The smaller birds are insectivorous, especially in the summer, although the bearded tit turns to reed seeds in cold weather. Insects occur in reed-beds in large numbers, especially the pyralid moths which lay their eggs inside reed stems. Bearded tits, and even blue tits, search the reeds for these larvae, often cracking open the stems to locate them. Many other wetland insects including mosquitoes and flies have larvae which are aquatic, the insects accumulating on the reeds as they emerge, providing a rich food supply for all the warblers, as well as swifts, swallows, martins and wagtails.

The prime conservation problem for wetland areas is to maintain the water level, both in the face of surrounding drainage systems, and in an effort to halt the natural evolution of a marsh to dry land. Continuous management in the past was supplied by reed-cutting for thatch. This is essential to prevent succession, since it allows new reed stems to grow the following year. A recent revival in the practice of thatching roofs has helped conservation organisations offset the cost of regular reed-cutting. Wetland reserves such as those at Wicken Fen and Woodwalton have managed to recreate extensive fen meadows, full of flowers, simply by cutting back the invading scrub and taking the area back one stage in plant succession.

149

Wetland Plants

As we have seen, the two types of wetland habitat which occur in Britain are the fens, found in lowland alkaline areas, and the true bogs which evolve where the soil is acid, usually in the hills. The fenland habitat can be divided into five obvious zones, each zone relating to the depth of water or to the degree of waterlogging of the soil which occurs, and each possessing characteristic plant species which favour those particular conditions. And although some species may overlap into neighbouring zones, certain wetland plants usually occur together, in characteristic communities. From dry land to the open water of the pond, these five main zones may be identified as follows:

1 Marsh zone – here the plants grow in waterlogged soil, but this is not usually flooded.
2 Swamp zone – characterised by tall, upright plants, like the common reed and bulrush, growing in shallow standing water.
3 Water margins or true emergent zone – where plants are rooted in the pond but their leaves and flowers are raised above it.
4 Submerged plant zone – here the plants are truly aquatic, living and reproducing beneath the water.
5 Free-floating zone – also true aquatic plants, but which float on the surface, obtaining maximum light for photosynthesis.

In many places not all these zones are present, succession depending upon altitude, water depth and acidity, and the vegetation can vary from a thin fringe of reeds or horsetails in an upland lake, to complete plant cover over the whole of a nutrient-rich lowland pond.

There are no truly aquatic trees in Britain, to compare, say with the swamp cypress of North America. However, the alder and several willow trees and shrubs do grow in very wet conditions in swamps and along freshwater margins, and influence the aquatic eco-system in three ways: they restrict the amount of light, their roots provide shelter for many aquatic animals, and their dead leaves add nutrients to the bottom of the pond or stream.

Some plants, like water mint and marsh marigold, are able to adapt to different conditions and thrive in both the marsh and swamp zones. Many of these marsh-zone plants have large, fleshy leaves so as to gain maximum benefit from direct photosynthesis – since water is always in abundant supply they have no need to conserve it, and so do not need small leaves as, for example, do plants on dunes or cliffs. The vegetation on mineral-rich soils is always very lush, but because the soil on the fens is often waterlogged it is thus deficient in oxygen, and so many marsh plants need broad, spongy or hollow stems to store and transmit air.

The marsh marigold is probably the best known marshland plant, producing a riot of golden flowers in early spring. Like most early flowers it provides abundant pollen and nectar for the many insects which have just emerged from hibernation. Purple loosestrife, with tall purple spikes, is a summertime flower which grows in large clumps by the waterside; so does the great hairy willow-herb, and both are an important source of food for bees and hoverflies. Another showy perennial is the yellow flag, its familiar yellow irises coming out in mid-summer; it, too, often develops into large waterside clumps which are then used frequently by coot and moorhen to conceal their nest sites. Water mint, with strongly creeping rootstocks, is particularly common in marshy ground, and so is meadowsweet, its flowers a lovely white cascade in summer; these last three species form a regular association on the marsh/swamp edge, together with yellow loosestrife, and blue tufted vetch and the pink flowers of ragged robin.

Other conspicuous flowering plants of the marsh include marsh St John's wort, bog arum, amphibious bistort, water forget-me-nots, marsh woundwort, monkey flower, giant water dock,

Fen flowers – yellow flag and ragged robin – are found in many wetland sites around the country. The flowers of the ragged robin are quite unmistakable giving rise to its name, while the yellow flag is very similar in form to the blue/mauve garden iris.

common comfrey, fen bedstraw, grass of Parnassus, marsh cinquefoil and meadow thistle. Typical fen orchids are the various species of marsh orchid, marsh helleborine and the very rare fen orchid. In the north, fringing the northern tarns, tufted loosestrife may sometimes be found; while in southern streams where the water flows more slowly great quantities of watercress may occur. In lime-rich streams there is brooklime, with its deep blue flowers, easily confused with water speedwell. These flowers are all perennials, often growing and reproducing through a creeping rhizome, rooting in the mud; many of them are found throughout the temperate zone of the northern hemisphere.

Most of these typical marsh plants develop into the swamp zone, although the plants characteristic to the swamp are taller and more upright. The swamp is covered with water in all but the driest years, and common reed, bulrush and reed sweet-grass are dominant, plants with few or no branches, long narrow leaves, and tough, fibrous, hollow stems all of which enable them to withstand wind and flood. Long, creeping rhizomes run through the mud helping to stabilise them, and new plants are reproduced from rhizome buds. In autumn, downy seed is produced in enormous quantities, especially from the reed and bulrush, dispersing on the wind to open water and so increasing colonisation.

Other common plants which grow regularly in standing water include flowering rush, conglomerate rush, branched bur-reed, club rush, great pond sedge and hard rush. All are tall plants with elongated, thin leaves. Sedges have rough, three-sided stems, whereas rushes have smooth, rounded stalks and leaves, often ending in a sharp point. Several species of both may grow regularly in the marsh/swamp edge and along the riverbank.

On the fringes of the swamp and towards the open water, emergent plants such as the water-lily are more typical. These are mostly true aquatics, thriving only when growing in permanent water – unlike swamp plants such as the reed which can survive long periods in dry conditions. Three somewhat similar fringing plants are the water horsetail, marsh horsetail and marestail; the horsetails are flowerless and produce spores in black terminal cones, while marestail is a true flowering plant bearing tiny, greenish petal-less flowers and equally tiny nut-like fruits.

There are two species of true water-lily in Britain, the white and the yellow; both live in the fairly deep marginal waters at the edge of a lake, their leaves providing shade, a resting place and cover for many aquatic animals. The yellow-fringed water-lily is similar, but is actually a member of the bog-bean family. Pondweeds, water crowfoot and water starworts all include a number of species which produce floating leaves and flowers.

Some of the crowfoots and starworts grow in still waters, but others are found in streams and rivers and the leaves of these are of a different shape and design according to the season. Thus, springtime leaves are fine and slender, because these cope best with the spates and strong water flow which occur in the spring; while summer leaves are broad, flat and floating so as to maximise the opportunity for photosynthesis.

Out in the open water are two further zones: the submerged plant zone, with fully aquatic plants such as willow-moss, milfoils, hornworts and Canadian pondweed; and the floating plant zone, typified by various duckweeds, water ferns and crystalworts.

Many of these plants are found throughout the northern hemisphere and some, like hornwort and crystalwort, occur throughout the world; this demonstrates the effectiveness of seed dispersal by water power, the seeds being carried for thousands of miles by rivers and sea currents.

The southern marsh orchid flowers in June in marshland localities of many southern counties. Its leaves are dark green and do not have the spots present on many orchid species.

Freshwater Birds

Water birds are often divided into two categories: open-water birds, and those of the freshwater marsh, reed-swamp and water margins. This is not a clear-cut distinction, however; wigeon, for example, may roost on the open water of a reservoir but feed on the grass of its banks, or a mute swan may feed on submerged plants in the middle of the lake yet make its nest in the surrounding reed-bed.

Upland lochs in Scotland have a distinctive group of breeding birds in summer. These include red- and black-throated divers – the former on the small hill-tarns – red-breasted merganser, goosander and occasional colonies of black-headed or common gulls. The diver populations are low and apparently still declining, with 1,200 pairs of red-throated divers and only 150 pairs of black-throated, mostly north and west of the Great Glen. Studies have shown that out of the total, Shetland has 101 pairs of the red-throated diver, and Orkney 45 pairs. Some lochs around Inverness have Slavonian grebes.

Most of these birds, with the exception of the grebe, are obliged to feed away from the food-deficient loch, sometimes even seeking fish out to sea. Upland rivers are also oligotrophic with little plant life, although flushes of caddis, stonefly and especially mayflies can be substantial, providing food for a few specialist fish – largely trout and lamprey – which in turn feed the mergansers and goosanders. Dipper, grey wagtail and common sandpiper feed along the banks, each exploiting a different food source, the dipper under the clear water surface for aquatic insects, the grey wagtail just above it for emergent insects, and the sandpiper picking at the shingle banks for invertebrates.

The bird life of these acid, upland bogs and meres hardly bears comparison with that of the rich, eutrophic, lowland lakes of the south and east; because the waters in these chalk or limestone areas have a high alkaline content, the level of nutrients is very much higher and so there are more plants, more invertebrates and of course, more birds.

Lowland rivers can support both the kingfisher and the little grebe (or dabchick); these are rarely found in the uplands. Both species are dependent upon the water for food, the kingfisher nesting in riverside banks and the dabchick in dense streamside vegetation. The kingfisher is a surprisingly small bird, only some 6½in (16.5cm) long, which catches small fish, sticklebacks and minnows by plunge-diving into the river. Cold, icy winters reduce its population dramatically. The more southerly distributed kingfisher numbers some 5–9,000 pairs, but it is quite rare as a breeding bird in Scotland. Another very common river bird is the moorhen, which conceals itself in the bankside vegetation to breed and raise its chicks.

Where a river or a pond has dense emergent vegetation along its banks the true residents of the reed-swamp will be seen, such as reed and sedge warblers. On larger ponds the dabchick is joined by the great crested grebe which prefers open areas for both feeding and display. This larger grebe feeds on fish supplemented by invertebrates such as dragonfly larvae both caught by diving. The legs of the great crested grebe are set far back on the body, an arrangement which is ideal for propulsion under water to catch its food but is ungainly on land; this is why it builds its nest in dense reed-beds, or attached to overhanging vegetation. All the diving ducks have their legs adapted in this way, too.

There are two diving ducks which are numerous on such waters, especially on the deeper sections and especially in winter: the pochard and the tufted duck. Both dive for their food, the pochard being mainly vegetarian while the tufted duck takes molluscs, and especially freshwater mussels. In winter, coot are also very common on these more open waters, diving for food in company with the two ducks.

The most numerous British breeding duck is the mallard, with perhaps 150,000 pairs, found on all suitable waters. It dabbles for food and eats both plants and invertebrates, hence its wide distribution. Once the breeding season is over the drakes

The kingfisher, not much larger than a sparrow, only catches small fish such as sticklebacks or bullheads which are beaten to death and swallowed head-first; it will also occasionally eat dragonflies. The kingfisher is adept at camouflaging itself – that is, until it flies when the brilliant blue of its plumage is a dazzling sight.

lose their brightly coloured plumage and by mid-summer resemble the drab brown females. This is called an 'eclipse' moult and they regain their bright plumage in winter.

Other 'dabbling' ducks which are commonly found on large lakes, especially in winter, include the shoveler and the much smaller teal, although the latter breeds in upland tarns and pools.

Several 'introduced' species occur, the Canada goose being the most widespread; it was introduced in the eighteenth century and it is thought that now there may be as many as 10,000 pairs, mainly in England – and its numbers are still increasing. The large, black-necked Canada is basi-cally sedentary, although one flock of 800 migrates from Yorkshire to the Beauly Firth in Inverness for its summer moult. Feral greylag geese are increasing too, from birds originally put down by wild-fowlers' associations.

The largest open-water birds are the swans. The bird of the local park lake is the familiar mute swan and it occurs over much of lowland Britain. However, its breeding population is in fact only some 5,000 pairs, widely and thinly spread. On angling waters it has been decimated by lead poisoning, due to its habit of taking up gravel with its food, when it also picks up the lead shot discarded or lost from anglers' weights – these are now illegal

A drake mallard in full spring plumage. By June he will have moulted into drab · brown 'eclipse' feathers, resembling the female's drab colouring with which she is camouflaged when sitting on her nest. The mallard's bill has a sieve-like mechanism enabling it to extract small animal and vegetable matter from shallow waters.

and the incidence of poisoning should decrease.

Winter brings two yellow-billed wild swans from the high Arctic, the whooper and the' smaller Bewick's swan. The Bewick adopts a more southerly winter distribution, its largest numbers at Slimbridge in Gloucestershire, and the Ouse Washes in Cambridgeshire. Whoopers prefer the lowland Scottish lochs. Some 6,000–7,000 whoopers and 16,000 Bewick's swans winter in Britain.

Large numbers of water fowl flock to Britain in winter, since our climate is mild when compared with the cold areas of the Continent. Many find refuge on reservoirs and gravel pits, both of which have proliferated over the past fifty years to provide modern society with water and building materials and have compensated for the drainage of many smaller, natural ponds. Wildfowl flocks are often larger on the eastern side of the country until there is a freeze-up, when many birds move south and west. Large numbers of coot, pochard, tufted duck and goldeneye, and also the rarer smew, scaup and goosander gather for the winter on reservoirs, along with thousands of mallard, teal and shoveler. Gulls, too, use these as safe night-time roosts, with over 200,000 coming to settle on the London area reservoirs alone.

The ducks, geese, grebes and swans which are normally associated with the open water of large lakes and ponds build their nests in the vegetation on its fringe, or in the ditches, dykes and tussocks of the open marsh. The surrounding marshland itself shelters many small perching birds (passerines) characteristic of this habitat, and also wading birds which do not use the open water at all.

The piebald lapwing is often the most noticeable bird of fresh-marsh fields; it has a distinctive moth-like flight, showing how it acquired its name, and hides its nest behind small clumps of rushes, the eggs typically blotched and well camouflaged like all wader eggs. There are invariably four with their pointed ends laid inwards. In spring their courtship displays are elaborate and include remarkable tumbling flight displays. Lapwings flock together in June after breeding, while large numbers arrive in Britain in October from the Continent.

The redshank and the snipe are the other characteristic fresh-marsh breeding waders, usually concealing their nests deep in tussocks of sedge and rush. As with all waders, the chicks move away from the nest within a few hours of hatching and feed themselves, preferring short, grazed grass. Both species display noisily in spring, the snipe in particular using a dramatic, switch-back 'drumming' flight in which the stiffened tail feathers produce the characteristic drumming 'bleat' each time the bird dives.

The yellow wagtail, the male resplendent in sulphur-coloured underparts, is also found in tussocky marsh fields. It is a summer visitor from Africa which arrives in late April, and nests in deep grass tussocks, catching insects from the vegetation. It is often found in company with grazing cattle, presumably catching the insects associated with them. Large flocks may roost in the reed-beds in autumn before migrating back to Africa.

Many of these marsh-bird populations are suffering a steady decline as their habitats are drained for agricultural purposes. The population of lapwings is estimated at 200,000 pairs, snipe at 30,000, redshank at 35,000 pairs and yellow wagtails at 80,000 pairs.

Reed-beds have also become a rare habitat due to modern drainage, but the large beds which do remain provide sanctuary for more specialist birds. Nearly all these birds are coloured in such a way that they blend completely with their surroundings. For example, the sedge warbler, which occupies the lower levels, has a striated buff and cream plumage which is the perfect camouflage; and so too does the bearded tit, a rarity whose range is restricted to reed-beds mainly in East Anglia. Both breed near the base of the reed stems, often in a tangle of vegetation. The reed warbler and the sedge warbler are summer visitors from Africa, but it is the reed warbler which is the characteristic songster of the reed-beds; pale buff above and white underneath, its colour allows it a measure of camouflage from above and below as it

sings from the upper reed stalks.

The bittern is extremely rare, and at one time it was feared it had disappeared from our reed-beds altogether. It has a vertically striped buff and black plumage and is, in fact, a large specialised heron. When frightened, the bittern will stand stock still with its striped neck held vertically – this successfully breaks up its outline and makes it look very like a bundle of reeds. Bitterns are now reduced to about thirty pairs, with the largest concentration at Leighton Moss RSPB Reserve in Lancashire where there are eight or nine pairs.

Herons too use fresh marsh ditches, where they catch fish and frogs. Most herons breed in woods overlooking the marsh, but where the reeds are extensive some may build their bulky nests in the tops of sallow thickets in the middle of the reed-beds for protection. The water rail is the classic skulking bird of the reeds, infrequently seen, but often heard squealing away in the depths of the beds. Icy weather may force it into the open as it looks for food. Coot and moorhen are much easier to see because they prefer open water and the reed-bed edges. They both swim well, though the coot is more aquatic having specially adapted lobed feet to aid propulsion, set far back on the body to aid movement through the water when diving. Moorhen are largely sedentary, but coot gather in large winter flocks and many thousands arrive in Britain to escape the continental cold. Winter coot numbers may reach 200,000 birds, including arrivals from as far away as Russia.

Perhaps one of the most surprising and numerous summer visitors to the reed-beds is the cuckoo. In May dozens may be seen over large reed-beds, busily searching for reed warblers' nests in which to lay their eggs.

There is only one raptor associated with large reed-beds: the rare marsh harrier, with perhaps only sixty pairs now found in Britain. It is a summer visitor from Africa, and preys on water birds such as the coot.

The heron is widely but thinly distributed throughout Britain and Ireland, with perhaps 9,500 pairs in total. It is particularly fond of eels, frogs and small fish which it stalks through the water with long deliberate strides. In flight its long neck is drawn back – not extended – and it flies with slow majestic wingbeats.

Winter brings dramatic changes to the marsh. Gone are the cuckoos and swallows, to be replaced by great flocks of duck, wigeon, teal, pintail and shoveler, out in the marsh fleets or cropping the grasses. With them may be large flocks of wild geese from the Arctic – white-fronted, pink-footed, greylag, barnacle, bean and Brent geese, sometimes in thousands. Fresh marshes in winter provide a continuous and dramatic wildlife spectacle of wide skies filled with birds.

Wetland Insects

Many species of insect use the waters of ponds and rivers during their life-cycle, especially in the larval or nymphal stages, but none is more noticeable than the dragonfly and its smaller cousin the damselfly. These may be seen to best advantage along the edges of a rich lowland pond in summer when the surface and vegetation will buzz with several species of different shapes and sizes. They are all quite harmless to humans, despite their country name of 'stingers'. In fact they cannot sting, and the sharp appendage on the tail of the males is a 'clasper' used in mating.

Dragonflies and damselflies belong to an insect order called the Odonata, the dragonflies grouped in the sub-order Anisoptera, and the smaller damselflies in the Zygoptera. They are all brilliantly coloured, often with a startling metallic sheen, and by comparison with their body size have enormous compound eyes which may contain over 30,000 individual lenses. Their predatory efficiency is further increased by the mobility of the head, which can swivel almost 360 degrees on the slender neck. Prey is caught in flight with the legs which are held together to form a 'net', it is then transferred to the powerful jaws.

Dragonflies can be divided into two groups: the 'hawkers', usually large, powerful fliers; and the 'darters', less powerful, slightly smaller insects which perch on the vegetation. The hawkers patrol a noticeable beat and display quite territorial behaviour, chasing other dragonflies away. They are all carnivorous, hunting and catching their prey in flight. The darters are so called because their

159

LEFT
A black sympetrum dragonfly has huge eyes to assist its hunting lifestyle. The larger dragonflies are known as hawkers from their habit of hawking around the edges of ponds; the medium-sized ones are darters and the small ones are damselflies.

RIGHT
The swallowtail butterfly is one of Britain's rarest species, found only in the East Anglian fens. The larvae feed on milk parsley, wild angelica, wild carrot and fennel. The mature caterpillar is as strikingly marked as the butterfly, with orange and black on bright green. Its breeding areas are designated nature reserves in an attempt to increase its population; it is being reintroduced to Wicken Fen in Cambridgeshire where it was once common.

method of hunting is to dart out from a favourite observation perch and catch any suitable passing insect, returning to the perch to consume it.

When a dragonfly is resting its wings are held out horizontally, or slightly depressed; the smaller, more slender damselflies rest with the wings held together over the back. These are equally colourful, and are also predatory, feeding on smaller insects; they sometimes fall prey themselves to the large hawker dragonflies.

For both dragonfly and damselfly, the greater part of the life-cycle is spent underwater as an egg or nymph; as the adult insect they will probably fly for only two or three months. When they are mating they will fly around 'in tandem', the male grasping the female by the back of the neck with his tail 'claspers'.

Some species lay eggs by piercing the stems of water plants, while others drop their eggs into weed-beds in the water; a dragonfly which is repeatedly dipping its abdomen into the water in flight is laying eggs. Summer-laid eggs hatch in a few weeks, while autumn eggs may overwinter. The nymphal stage may last from one to five years, depending upon the species and the climate. Damselfly nymphs in a pond are easily recognised by the three leaf-like gills at the end of the abdomen. Dragonfly nymphs have an internal gill structure situated in their rear end.

The nymphs are carnivorous too, but must rely on stealth and camouflage to catch their prey underwater; what they eat ranges in size from water fleas to small fish, and cannibalism occurs frequently. The nymph waits for a victim to pass within reach of its 'face mask' which is highly flexible and which it can extend in a flash, impaling the prey on its terminal hooks.

Adult dragonflies and damselflies emerge overnight or at dawn. The nymphs crawl up the vegetation at the water's edge so as to get above the surface, and split open, allowing the adult to crawl out, rather in the way of butterflies. The young adults are soft and very vulnerable until their wings harden and they take flight.

161

Emergence takes much longer than it does for a butterfly, and it may be twelve hours before the insect flies.

There are forty or so species in Britain, although three or four may have become extinct in recent years because of pollution. Most species need unpolluted, still water, but a few are found on faster-flowing streams. Two damselflies in particular prefer flowing water: the metallic blue/green *Calopteryx virgo* and *splendens*, and the rarer white-legged damselfly; and also two larger dragonflies, the club-tailed and the golden-ringed.

The first dragonflies appear in early May and most are flying by mid-June, although a few species such as the brown aeshna, the common aeshna and the black sympetrum appear between midsummer and late summer. Over still waters the most frequent species are the common blue and large red damselfly, the four-spotted and broad-bodied libellulas and the southern aeshna. Five species which breed regularly in Britain are migrants, and there are sometimes huge movements of dragonflies in south coast localities like Dungeness and Selsey Bill.

Most of the other aquatic or waterside insects are not as visually striking as the dragonflies, except for two brilliantly coloured butterflies which are found only in fens and marshes: the swallowtail, and the large copper. The swallowtail butterfly is a rare insect found only in favoured localities in the East Anglian fens and broads. Over here its larvae feed solely on marsh hog's fennel, or milk parsley, although on the Continent a wider range of umbellifers is used. It is on the wing in late May and June, with a partial second brood late in August. The striking scarlet-coloured large copper is confined to one small fen, where its population is maintained by special protective measures brought into force for its breeding season, mainly to ensure its food plant, the great water dock. It may be seen in July.

Among other wetland insects are mayflies which may appear in thousands in early summer. They are easy to identify, with large triangular forewings and small hindwings criss-crossed with a pattern of veins like a net. Their adult life is very short as the insects do not feed; the mayfly flight is solely concerned with reproduction. There are also caddis-flies which build their own home, fashioning tiny bits of wood and stones into a tube-like structure which they then occupy. And stoneflies, clinging to stones in the upper river.

The pond skater is an easily recognised, fast-moving bug which lives on the surface of ponds, ditches and slow rivers. They are often found in great numbers and have predatory habits, catching insects and other small animals trapped in the water surface – they themselves can skim quickly along the top as their legs have a pad of bristles which supports them on the surface tension film of the water.

Water beetles, water boatmen and back-swimmers are all insects which live in the water and are therefore truly aquatic. They feed on other insects and animals and can fly well, and will often migrate from one pond to another.

Freshwater Fish

There are over fifty species of freshwater fish indigenous to Britain, and many more introduced species. How these would be distributed if left in their natural state is difficult to ascertain, because so many fish are transported by man from one water to another with a view to his own sporting interests and for commercial production.

Most fish in Britain are found in rivers, lowland lakes and ponds. With one or two exceptions, fish are rare at high altitudes in the mountains, especially as the waters of hill-lakes tend to be peaty and acid, and do not support a good food supply. Nor will fish survive if the water suffers from a high level of pollution, or a low level of oxygen.

In its upper reaches a river is usually fast-flowing, turbulent and full of oxygen and is inhabited by those renowned game fish, the salmon and the trout. The salmon is really a marine fish which adapts to freshwater because it comes upriver to breed, laying its eggs on gravel beds in the upper reaches. Salmon may leave the sea and enter the river at any time of the year, and when they do so varies from river to river. Fish which enter before November usually spawn by Christmas, while those arriving later may spend a whole year in the river before spawning. Many die after the effort of reaching the spawning grounds, although some do survive to make a 'run' again. Young salmon are known as 'parr', and will remain in the river system for several years before returning to the sea.

The sea trout and the brown trout are the other fish typical of fast, well-oxygenated rivers. The sea trout is often difficult to tell apart from a salmon, and is also a marine fish which spawns in freshwater; the brown trout, however, is a full freshwater species. It spawns in autumn and winter in gravel beds in running water, and the fry hatch in spring. Upland lakes and rivers do not have a good food supply but where conditions are favourable the young fish may reach 6in (15cm) by their first autumn. A weight of 4–6lb (2–3kg) would be high for a brown trout and could take seven years to attain.

The other great game fish is the grayling, which

basically shares the same habitat as the trout. It is especially common in the chalk streams of Dorset, Hampshire and the Severn Valley.

Besides the game fish there are other, smaller fish which occupy the same waters, and which also need a good supply of oxygen to survive; they include the minnow, found in shoals in clean, clear water (and usually in the shallows to avoid being eaten by larger fish), the stone loach, and the miller's thumb, or bull head. The last two are both fish which inhabit the gravel and pebbles of the river bed; all three spawn in spring.

The middle reaches of the river support barbel, chub, dace and bleak: barbel feed on the river bottom in relatively fast-flowing stretches; chub live in similar areas but swim nearer the surface; and dace occupy faster, shallower water, preferring rivers with a clean gravel bed.

The fish species which thrive in lowland rivers with slow currents and many submerged water plants are similar to those which are found in weed-filled lakes and ponds, and far more fish occur in these lowland, food-filled still waters than in the hills. The species which would normally be found in these eutrophic waters are roach, rudd,

The rudd is one of the most numerous fishes found in the slower-flowing reaches of rivers. It is similar to the roach, but has more rufous fins. The rudd is noted for shoaling and hybridising with other closely related species such as roach, bleak, bream and white bream.

bream, perch, pike, carp, tench and orfe; these are known as 'coarse' fish (as opposed to 'game' fish – the salmon and trout), and they breed in the dense beds of aquatic plants. The fish which feed on small organisms tend to form shoals, and are active all the time in their search for food. These are mid-water fish and usually have large eyes; tench, barbel and carp are bottom feeders and have small eyes; they 'feel' for their food in the muddy bottom with the barbels around their mouth.

The larger predatory fish such as the pike are territorial and solitary, and lurk in the shadows waiting for likely prey to approach before making a quick dash out to seize it. The pike is the predator *par excellence* of the lake and may grow to six feet (two metres) in length and weigh 50lb (22.5kg). Fish of this size can easily swallow ducklings and other small water birds.

The appearance and colour marking of different fish species often provides a good indication as to their habitat; for example, perch have vertical stripes and usually live among the weed stems, while the tench, a fish of muddy bottoms, is a muddy brown. Perch are considered as predators but behave differently from the others since they normally live in shoals.

There are two types of stickleback, the three- and the ten-spined and the behaviour of both changes according to the time of year. Outside the breeding season they live unaggressively in large shoals; however, during the breeding season the tiny 'stickles' are as territorial as the pike, the male fiercely guarding its nest in the weeds and caring jealously for its young.

Eels though migrants from the sea can reach isolated ponds because of their ability to move long distances overland through damp vegetation, breathing through their gill-chambers.

Fish feed on the aquatic flora and fauna; some are carnivorous like the pike, while others, such as roach and rudd, feed on plants and insects. They can all survive without food for long periods – especially useful through cold winter weather – since they lay down a store of fat during the summer. And Britain's fish population is also well adapted to the climate, with one species or another breeding in every month of the year: dace in January and February, grayling in March, through to salmon in November and December.

Frogs, Toads and Newts

There are six native species of amphibian in Britain: the common frog, the common and natterjack toads, and the common, palmate and great crested newts. In addition there are one or two foreign species which have been recently introduced in the south. Worldwide there are estimated to be over 4,000 species of amphibian altogether.

All six British species hibernate through the winter. The best time to see them is in early spring when they return to their traditional breeding ponds; it is essential for them to find water in which to breed because they lay soft eggs without any protective outer shell. Water is also essential because like most amphibians, they breathe

Pike are ferocious predators of slow and still waters, taking other fish, water-birds such as young moorhens and ducklings, and also water voles. The pike moves with incredible speed, covering ten times it body length in a second, and is armed with needle-sharp teeth from which its prey stands little chance of escape.

through their skin, and this must be kept moist to allow easy passage of oxygen into, and carbon dioxide out of the body; mucous glands in the skin assist in this task. Some also have lungs so they can obtain oxygen when they are out of water, in the way of most mammals and birds; this capacity is common to the British species.

The common frog is a widespread species except in areas of arable farmland, where modern farming methods have decimated numbers. Suburban gardens with ponds, on the other hand, provide frogs with an ideal and secure habitat, although domestic cats can be a major hazard.

The common toad is more of a lowland dweller than the frog – frogs have been recorded as living in small mountain tarns at altitudes of over 3,000ft (1,000m). The toad can tolerate dry conditions better than the frog and spends nearly all its life away from water; frogs need damp, marshy condi-

tions particularly after they have left their breeding ponds.

Frogs and toads are easy to tell apart: the frog is smooth-skinned and slim with a pointed nose, and a greeny-brown colour with a distinctive brown patch behind each eye; while the toad is sturdy and thickset, with a dark mud-brown skin and a blunt, flattened nose.

Frogs arrive in their breeding ponds a little earlier than toads, sometimes by late January or early February. Toads in particular make long migrations, sometimes even as much as two or three miles (3–5km), across country to reach their own pond, arriving by early March. They often have to cross roads where these are comparatively recent and intersect with the older, traditional toad-migration routes. In this respect, the British attitude to conservation is often poor by comparison with our European neighbours; in France, Germany and

other northern European countries some roads may be closed to traffic at toad migration time. Britain relies on 'toad on road' signs but drivers often pay little heed to these, with devastating effects.

When travelling on land toads waddle or crawl along, but frogs hop. Once they arrive at their respective ponds the mating activity is frantic as the males struggle and fight to secure a female. The male is carried around on the back of the larger, spawn-filled female until the spawn is laid and fertilised. Frogs usually lay their eggs in shallower water than toads, in great jelly-like masses resembling tapioca pudding. The toad lays long strings of eggs which may be 10ft (3m) in length and contain 3,000 eggs. Natterjacks, however, lay short strings in shallow water.

Enormous numbers of eggs are produced because most of the tadpoles will die, only a few surviving to adulthood. Cold and fungal infection kill many of the eggs themselves, and when they be-

come tadpoles they are eaten by a wide variety of predators from diving beetles to herons. Toad tadpoles – 'toadpoles' – are jet black, whereas frog tadpoles – 'frogpoles' – are brown with golden speckles. Natterjack tadpoles are very like those of the common toad, but are usually found in pools in sand dunes.

Tadpoles eat algae and other small green plants, but will also eat dead animals and occasionally each other. By midsummer both froglets and toadlets will have gained their legs, and will leave their ponds as tiny replicas of their parents; they take some two to three years to mature.

Many predators eat frogs but a toad has two defence mechanisms: it can puff itself up until it is some 30 per cent larger in size, which may persuade a grass snake that it is too big to swallow; and it can also secrete powerful poisons from its warty skin – these include several alkaloids such as bufotenine and bufotenidine, which make it very unpalatable. You should wash your hands after handling a toad.

The natterjack toad is now quite rare. It is an animal of southern Europe, with Britain right on the edge of its range, and is found in the sandy coastal dunes of southern and eastern England and on the dune coasts of Lancashire and the Solway. A few colonies also exist in south-west Ireland. The natterjack is a small green toad with a bright yellow line down the middle of its back. They are agile creatures and very noisy in the spring breeding season, which starts in April as the weather improves.

Several other species are recent introductions. For example, the marsh frog, common on the Continent, was released on Romney Marsh in 1935, when a handful were put out in a garden pond. They thrived, and eventually occupied all the ditches and ponds in the district. These are large frogs and the males indulge considerably in croaking displays in spring. However, the very cold winters of the early eighties depleted the population considerably.

In the Home Counties there are several colonies of the edible frog, introduced mainly in the last century. Most are small, but there is a colony in one old Surrey claypit which is really quite large, and many hundreds appear each spring. Midwife toads and European tree frogs have also been introduced, and survive in one or two favourable localities, usually as an isolated community in one sheltered pond.

Like frogs and toads, newts are readily found in spring in the ponds where they travel to breed. They spend the rest of the year hiding under stones, fallen wood and grass tussocks, venturing out mainly at night to eat worms, slugs and insects. The smooth or common newt is probably the most numerous of the three British species, but is mainly restricted to alkaline waters, mostly in the south; the palmate newt is the smallest and prefers more acid waters, and is therefore commoner in the north and west of Britain. The great crested newt is now rare in Britain, found only in scattered colonies in England and Wales. It is quite large,

LEFT
Three pairs of mating common toads in early spring. Toads are protected from predators by their vile taste and their surprising ability to puff themselves up in size. Toads lay their spawn in strings, whereas frogspawn is laid in large jelly-like masses.

RIGHT
The great crested newt is rare and has been given special protection by the Wildlife and Countryside Acts. It has a distinctively warty skin and, like the common toad, is highly distasteful to predators. The smooth newt, the commonest of Britain's three species, is widespread; it lives in ponds in spring and summer but hibernates on land from mid-October to February.

about 6in (15cm) long, and like the toad, can produce poisons from its warty skin, presumably to protect it from being eaten. Newts may be eaten by grass snakes, herons and predatory fish; and in 1987 a vagrant little bittern demolished a whole population of legally protected great crested newts when it stayed at an East Sussex pond!

The female newt is always drab in appearance, but the male develops bright colours in the spring to enhance his courtship displays, and the great crested and the smooth newt gain a notable crest. The male crested newt has a bright orange and black belly, while the male smooth newt has a red underside dappled with black spots, and a blue line along the bottom of the tail. The male palmate newt is green and gold with black, webbed hind feet. The breeding procedure is unusual, as the male deposits a sperm packet – a 'spermatophore' – on the pond bed during a complex courtship and this is picked up by the female in her vent. Fertilisation of the eggs is internal as opposed to the external fertilisation of frog and toad eggs. The tadpoles are tiny and take until August to develop. Young newts take two to three years to become fully mature, and do not return to their pond until then.

Wetland Mammals

Only a few of Britain's mammals are characteristic of wetland habitats. These are the otter, water shrew, water vole, mink, coypu and Chinese water deer. The last three are either escapees or introductions; and the coypu, once widespread in East Anglia, is probably extinct, having succumbed to recent very cold winters and an all-out campaign to eradicate it.

The classic wetland mammal is the otter, included here as a freshwater mammal although its

The wild otter is now rare in Britain; it is most often seen on the coasts of the north and west of Scotland and in Ireland. A fish-eater, it has been persecuted by fishermen and has also been badly affected by pollution. It is distinguishable from the mink by its larger size, white throat and longer tail.

sad decline in Britain means that it is now numerous only on the sea coasts of the northern islands. It has virtually disappeared from lowland southern Britain and now only survives in any numbers in Wales, the Highlands and in particular on the sea-coasts of the Scottish islands and the north-west mainland. It is very wary, and now rare, having declined dramatically in the past thirty years. Some blame otter hunting, which is now banned, but in fact it was the concern of the hunts themselves and their analysis of hunt records which alerted people to the serious decline in population numbers. The main reasons for the otter's demise are pesticides, water pollution, the clearing and damming of rivers, and disturbance, especially that caused by the vast increase in recreational activities on all waters.

Otters are slim and lithe, about 3–4ft (1m) from nose to tail. The male – the 'dog' otter – is larger than the female – the 'bitch' otter – and averages about 24lb (10kg). Both have a chocolate-coloured glossy coat with pale fur on the underside. This coat consists of two layers: a very dense, fine underfur, with a thick topcoat of long, oiled guard hairs which together provide a warm, waterproof and insulating covering. Nevertheless an otter rarely spends more than half an hour in the water as it becomes chilled, particularly in northern waters, and is prone to pneumonia. It will emerge and go through an elaborate grooming process to dry its fur and warm up.

Fish comprise about 80 per cent of the otter's diet, but it will also eat frogs, crayfish and some invertebrates, and is especially partial to the 'long' fishes such as butterfish and blennies in the sea, and eels in freshwater. An otter den is called a

'holt' and is usually situated under riverside tree roots or boulders.

In the wild, otters seem able to reproduce at any time of the year – perhaps fish are easier to catch in winter when they are cold, and thus a spring/summer breeding season is not necessary. The bitch is an exemplary mother, rearing two or three cubs after a short gestation period of some nine weeks. The cubs stay with the mother at least till the following spring so she may not breed every year.

Otters are more nocturnal than diurnal, although if they are not disturbed they may be out and about for quite some time during the daytime too. To contact each other they utter a loud whistle, which is often one of the first indications that an otter is nearby. They will mark out their territory by regularly depositing their droppings, known as 'spraints', on prominent objects such as logs, rocks and tussocks along the riverbanks. Spraints are black and oily, and fish bones can often be seen in them.

One of the most notable escapees in this country is the American mink, originally farmed for its fur and now widespread throughout Britain. The feral mink usually has very dark brown fur with a white throat patch although other pastel shades turn up. It is related to the otter but is much smaller, averaging about 2lb (1kg) in size and some 2ft (60cm) in length. It has become one of the most numerous of Britain's carnivores, even though its range is restricted to wetland areas. Early fears that it would decimate fish and water-bird stocks appear generally to have been unfounded and the mink seems to have settled into its waterside niche. Some 70 per cent of its diet consists of small waterside mammals but it is an adaptable feeder, and supplements this with fish, crustacea, birds and even beetles and worms. Mink are quiet and secretive, but are most likely to be seen on streams; they may be located by their footprints

The Chinese water deer is now an animal of the fens, though the British population has actually grown out of a number of past escapes from Woburn Park. In size it is little larger than a labrador.

and foul-smelling black spraints.

Chinese water deer are small, shy, marshland deer which are found in a few quiet sites largely in East Anglia; they originally escaped from Woburn Park and are most likely to be seen as a chance encounter in one of the Fenland nature reserves. The males have tusks rather than antlers and the rut occurs in winter; fawns are born in May or June, and unusually for a deer there are often twins, triplets or even quads.

The two smallest wetland mammals are also the most numerous: the water shrew and the water vole. The water shrew lives in dense vegetation and is by no means confined to waterways. It is a pretty little animal, black above, white below, and like its close relatives, the common and pygmy shrews of grassland, it needs to eat constantly to maintain its high metabolic rate. It is a good swimmer, using its fringed hind feet to propel it through the water, and eats invertebrates such as dragonfly larvae. It can also kill fish and frogs and, like other shrews, its saliva may well be toxic, giving it a poisonous bite.

The water shrew is found largely on the mainland of Britain; it is absent from Ireland and most of the islands. It is small, weighing about ½oz (12–18g) and moves about more than other shrews, with a range of up to 220yd (200m). It breeds from April to September but normally has only two litters with an average of about seven young per litter. These are weaned at twenty-eight days.

The easiest of all wetland mammals to locate is the endearing little water vole, misnamed the 'water rat'. The water vole leads an easy-going, vegetarian life along the riverbank, preferring quiet, slow, streams with plenty of streamside vegetation. It is not found in Ireland, but it does occur in Scottish burns up to 3,000ft (930m) in altitude. Its weight is variable, being greatest in spring when it may reach up to ½lb (200–300g). Its home range stretches as much as 220yd (200m) along the river bank. Sadly, the water vole population has recently suffered a sharp decline, probably as a result of water pollution. The female produces four to five litters a year, with four to five young in the average litter.

Water voles are inquisitive animals, and quite diurnal, easy to see if you sit quietly on the bank of a stream.

CHAPTER 7

ESTUARIES AND LOW-LYING COASTS

ESTUARIES and salt marshes are often considered by urban man as wasteland, just so many unproductive acres ripe for 'development'. However, they are in truth very productive habitats, an extremely rich source of food for millions of shore birds and a veritable production centre for young fish and crustacea, including many species which man exploits commercially. It is interesting to reflect that salt marsh produces some four times more weight of vegetation than agricultural cereal fields.

The estuarine eco-system owes its very existence to the rise and fall of the tides. A high and a low tide occur once every twelve hours or so, and at least once a month there is a 'spring' tide, when the sun and moon are nearly in a straight line and the extra gravitational pull produces a maximum effect of ebb and flow. Very high spring tides occur in September and March at the equinoxes, for the same reason. 'Neap' tides are those where the ebb and flow pattern is at its lowest.

In Britain the coastline extends to approximately 9,000 miles (14,500km) in length. All around this coast there are innumerable estuaries and bays, but of particular interest are the 133 or so which each possess ½sq mile (1sq km) or more of tidal flats, totalling some 1,000sq miles (2,600sq km) altogether. This may sound a lot, but the Waddensea on the German north coast covers nearly this area alone. The estuaries and tidal flats represent a very precious wildlife resource, especially for birds in winter; national wildfowl counts have shown that over 2 million waders and 1 million wildfowl use them, presumably because they find them relatively undisturbed and can feed and rest there in peace. These figures represent some 30 per cent of the whole European populations of these birds, so even if our islands are small, Britain is a vitally important winter refuge for hordes of birds from many other countries.

Most of the estuaries in Britain were shaped by the gradual rise in sea level as the ice in the last Ice Age melted. The sheltered conditions also mean that a river's erosion products, its silts and detritus, will be deposited along with sands and pebbles brought in by the tide. This produces the estuary's most striking feature, its vast inter-tidal flats, full of mineral salts and supporting huge populations of invertebrate animals. Animals living here must adapt to a cycle of constantly changing salinity, water level and temperature, and a regular renewal, by the tides and the river, of energy and nutrients; they will also run the risk of dehydration at low tide in hot weather. Other important factors affecting the resident wildlife are the type of bottom or substratum, the current and wave strength, and the amount of dissolved oxygen. Particle size is of particular importance; where the silt is too fine many invertebrates cannot survive because their breathing systems become clogged. Moreover, in fine silts bacteria produce the poisonous gas hydrogen sulphide and this also prevents colonisation by the estuarine animals – with the exception of tubifex worms which have adapted to survive there.

The large particles tend to drop from the tide-race early on, which is why shingle banks build up at the estuary mouth. The mix of mud and sand with particles of intermediate size provides the best conditions for invertebrates, with plenty of space for oxygen and food in the form of plant detritus between the particles. When the invertebrate populations living in these tidal flats have been assessed, the densities per square metre have been impressive: laver spire shells, a tiny gastropod some ¼in (6mm) long, reaching 35,000 per square metre; the tiny bivalve macoma in excess of 50,000 per square metre; a small amphipod about ⅓in (8mm) long, Corophium, also reaches these densities; and ragworms, which despite being much larger, can still achieve 1,000 per square metre. Clearly these do not all live in the same section of mudflat at the same time; nevertheless, 50,000 items per square metre is not an unusual estimate in the most productive estuaries.

Sandy flats may still have considerable numbers of invertebrates, especially bivalves such as cockles, provided the supply of detritus food, and minerals, is maintained, although the numbers are markedly less than in a mud and sand mix. Bird populations will also be lower because there are fewer prey items. The detritus brought down by rivers supplies a part of their diet, but their primary source of food on estuaries and low-lying coasts is the salt marsh itself which builds up in the more sheltered parts of the estuary. Salt marshes are the powerhouse of the system – yet it is precisely these marshlands which are considered by developers and politicians alike to be waste ground, an environmental mistake which could have disastrous ecological consequences.

Salt-Marsh Plants

There are a few plants which have colonised the estuarine environment successfully, accepting the periodic submersion by salt or brackish water, and these fall into two clear categories. First are those plants which will tolerate being covered regularly by all tides; these include several algae, especially the various enteromorpha species which grow as long tube-like strands of green vegetation on mudbanks; and one genus of true flowering plant, the eel-grasses, or Zostera. Both constitute an important food for ducks and Brent geese.

The second category are maritime, as opposed to marine, plants and the first of these to take hold in the bare mud are usually glassworts – Salicornia where the tides' currents are weak, and sometimes maybe cord-grass, Spartina, or the grass Puccinellia which has stronger roots. More mud is then trapped by the roots of these first colonists, and with the annual cycle of growth and regression more dead plant material and yet more mud steadily accumulates. This steady accumulation of material raises the salt-marsh level so that as the years pass, it is flooded less often by the tides. Measurements on the north Norfolk coast of Scolt Head have shown accretion rates of ½in (1cm) of mud per year, and levels up to 8in (20cm) per year have been recorded in France. Other, less salt-tolerant plants become established on the higher levels as inundation by the tides becomes less frequent, and thus the marsh continues to rise until only the very highest tides may affect the inner areas. This is the reason for the noticeable zonation of plants on the salt marsh. Obtaining fresh water is their prime problem because although water is abundant they cannot use it owing to its

PREVIOUS PAGE
The sandy estuary of the Taw and Torridge rivers in North Devon provides an ideal habitat for large flocks of waders, especially oystercatchers. Plover, lapwing, dunlin, curlew and redshank are also seen in large numbers.

RIGHT
Common sea-lavender is a typical plant of salt marshes, flowering from July to September. Another member of the sea-lavender family is rock sea-lavender which flowers on coastal cliffs and rocks as well as, occasionally, shingle and sand.

high salt content. Strong drying winds, a feature of low-lying coasts, adds to the problem. Many salt marsh plants have adaptations to reduce water loss such as hairy leaves (sea purslane), waxy surfaces (sea holly) and thick outer skins. Some, such as glasswort, are succulent, storing water in their leaves.

Glasswort, also known as marsh samphire, is always distributed on the extremities of a developing salt marsh. On the seaward edge, isolated plants manage to take hold in the bare mud, their density increasing shorewards, and then intermixing with colonies of rice-grass. This results in an increasingly stable accumulation of mud further inshore which in turn produces conditions favourable for stands of sea-aster and annual sea-blite.

A little higher, and a little drier, and the marsh is dominated by sea-lavender, filling the area in late summer with its purple-blue flowers. A little drier still and the marsh is covered in sea-pink, or thrift, which may cover acres of foreshore; for example, as at Lindisfarne in Northumberland. The top of the marsh may be dominated by red fescue-grass, sea-rush and a variety of oraches.

A salt marsh builds up around clumps of vegetation, the tide finding a passage for itself round the areas of more firmly anchored mud. In large areas of marsh these drainage creeks may form an intricate pattern of ditches and runnels which are often half covered by vegetation. A walk across a salt marsh should therefore be undertaken with great care, though the creeks are often lined with low scrubby bushes of the grey-green sea-purslane – being able to recognise these will often save a plunge into a 3ft (1m) deep, muddy runnel.

The cord-grass, spartina, was brought over from America in the 1870s, probably in ship's ballast. It cross-fertilized with a British species creating a brand new and aggressively colonizing hybrid S. *townsendii*. This species has spread all around our coasts and estuaries and has caused great problems, not least because it is so invasive that it does not allow a normal salt-marsh plant succession; in the oldest areas this has resulted in extensive die-back, leaving a poor soil exposed to rapid erosion. It also destroys the invertebrate fauna and therefore has long-term detrimental effects on bird populations.

Estuary Birds

As the seasons change so do the birds which are to be seen on the estuary, and nearly all of them are migratory, travelling enormous distances across the globe. Even the apparently sedentary mallard which breeds in the marsh in summer may winter in Spain, and the mallard feeding in winter may have come from Russia.

There are specialist breeding birds, passage migrants, and, in winter, hordes of birds from the high Arctic, and this constant bird traffic in and out of the estuary gives it a touch of excitement at all times of the year. It is no wonder that most of Britain's one million birdwatchers head regularly for the estuaries. Estuary birds, both waders and wildfowl (ducks, geese and swans) have learned to live with the constantly changing environment, each species exploiting a separate niche in its search for different food items. Many species depend on the huge numbers of invertebrates living in the tidal flats.

A group of specialists breeds on the estuaries themselves, while many more species prefer to be just over the sea-wall on the adjacent fresh marshes. Fresh marshes were once used for grazing cattle and are a vital part of many estuarine systems, but many have been ploughed up and turned into wheat fields. The two habitats are usually divided by nothing more than a man-made sea-wall and birds from one area very often use the other for feeding and resting.

Much of the information concerning the numbers of birds using estuaries comes from the Birds of Estuaries Enquiry set up in 1969 by the British Trust for Ornithology, aided by the Royal Society for the Protection of Birds, the Nature Conservancy Council and later on the Wildfowl Trust. Although the survey ceased in 1975, wildfowl and

A common tern at its nest in a scrape in the sand. Unlike the Arctic tern it has a black tip to its bill, although the difficulty of distinguishing the two species has led to them both being popularly known as 'comic' terns.

wader counts are still regularly undertaken throughout each year by an army of skilled amateur ornithologists.

The breeding season starts early in spring, at a time when many migrant waders and wildfowl are still moving north. Shelduck will be searching for suitable rabbit burrows to breed in, on the highest part of the salt marsh; mallard nest among the dense vegetation along with many redshank, using cover which is really quite deep when compared with the isolated grass tussocks they would use on the open fresh marsh. Breeding passerines include skylark, meadow pipit, reed bunting and linnets, which often use the taller sea-purslane scrub. Here and there around the coast are colonies of black-headed gulls, common terns and sandwich terns, all of which use the higher salt marshes. These last three species may well have spilled over on to shingle and sand spits, and large colonies may build up in favourable localities, above high-water mark and free from human disturbance. These shingle or shell beaches often have a few pairs of ringed plover and oystercatcher, and the little tern breeds on these beaches too, although it is highly vulnerable to trampling human feet. Recently intensive wardening has kept the public at a distance, and this has helped small colonies produce more chicks, but the little tern is still one of Britain's rarest breeding sea-birds.

The population of ringed plovers is estimated at

8,500 pairs. They produce two broods of three to four eggs, incubation takes about twenty-five days and fledging a further twenty-five, although the chicks leave the nest immediately on hatching. The oystercatcher population is estimated at about 35,000 pairs. It produces one brood of two to four chicks, incubation takes about twenty-six days and fledging five weeks. The common, little and sandwich terns all have one brood, usually of one to three chicks, incubation taking about twenty-three to twenty-five days and fledging four to six weeks. The young leave the nest within a day or two of hatching. Current population estimates are 12,000 pairs of common terns, 14,000 pairs of Sandwich terns and 2,400 pairs of little terns.

The autumn migration of estuary birds starts early and by mid-July large flocks of 'new' black-headed and common gulls have arrived while local birds are still busy at their colonies. Waders reappear, often 'failed breeders' departing early from the far north and still in colourful breeding plum-

Oystercatchers at a high tide roost. The oystercatcher's dagger of a bill is used for opening the valves of mussels and smaller bivalves. Extremely conspicuous with its black-and-white plumage, pink legs, red eyes and bill, it is a noisy and excitable bird and its long shrill 'flee-eep, klee-eep' call can be heard over long distances.

age – grey plover and dunlin with black chests, and a few black and bar-tailed godwit in bright red plumage. The July trickle of migrants soon becomes a flood, with peak numbers moving south in the last two weeks of August. A wide variety of species all feed on the mudflats and in the creeks, with large flocks of curlew, lapwing, oystercatcher, redshank, ringed plover, grey plover and many more. A huge build up in hirundines occurs from late July onwards, with waves of swallows and sand martins flitting south along the shore, feeding on the marshland insects.

September sees the first influx of wintering

wildfowl, with teal, wigeon, mallard and shoveler arriving from the Continent. A further wave occurs in November, and a final influx round about Christmas time, the birds preferring to escape the grip of the cold continental climate. Wild geese also arrive from mid-September onwards, the first barnacle geese from Spitzbergen appearing on the Solway at about the same time as the pinkfeet from Iceland would be pitching in to Martin Mere in Lancashire. By late October the Brent geese will have arrived on the south and east coasts; and finally the white-fronts, which are often the last to appear, in December, on their traditional feeding grounds of the Thames and Severn estuaries.

Winter is the prime time for birds on Britain's estuaries. Although we may not always appreciate it, Britain is relatively mild in winter compared with much of the Continent. November sees a dramatic influx of birds which have spent the autumn on the Waddensea on the Dutch and German coasts. Dunlin, knot, bar-tailed godwit, shelduck, wigeon, Brent geese and many more increase in number until they reach a peak in population by about mid-winter, so at the turn of the year Britain's estuaries are at their most crowded. The best time to see these birds is on a rising tide when water covers their inter-tidal feeding grounds and they are forced on to higher ground to wait out the tide. These high-tide roosts may hold many thousands of birds in favourable localities. Totals for Britain from the wildfowl and wader counts are impressive, with half a million dunlin in top place, followed by 230,000 knot, 200,000 each of oystercatcher and golden plover, 100,000 curlew, 80,000 redshank, 44,000 bar-tailed godwit down to 350 spotted redshank and avocets. For wildfowl, top place goes to mallard with 300,000, followed by 200,000 wigeon and teal and 100,000 pink-footed geese. A few birds of prey come to hunt these large bird flocks in winter – peregrine, merlin and hen harrier occur in small numbers on the larger estuaries.

By March these huge flocks are dwindling as birds return to their breeding grounds as far afield as Siberia, Scandinavia, Iceland, Greenland and northern Canada. Throughout March and April a stream of replacements pass through as those that have wintered farther south, in West and South Africa, continue northwards.

Feeding Methods of Estuary Birds

The estuaries of Britain are vital feeding areas for hundreds of thousands of birds, especially in winter. Their diet depends on the myriads of invertebrates which live in the inter-tidal flats, a habitat more productive than the most intensively farmed land.

Waders tend to concentrate on particular items of food, and to a large extent the size and the choice of the food item depend on the depth to which it burrows and the bill size of the bird concerned. Therefore grey and ringed plovers which have the shortest bills feed mainly by pecking the hydrobia – spire shells – and other species living on the surface. These spire shells, together with one or two other extremely numerous invertebrates, are important in the diet of many waders. Birds which feed on the surface adopt a 'watch and run' feeding technique, bobbing down to pick up a titbit of food as it appears, then running off along the mud until they see the next. The dunlin has a longer bill, of about 1¼in (3cm), and can therefore probe the thin top layer of mud which is inhabited by many species and especially the amphipod, Corophium. Packs of knot with bills of similar length to the dunlin probe down some 1½in (3–4cm) for the Baltic tellin, a tiny shellfish related to the cockle.

Other small waders eat a diversity of other food items which originate in various nooks and crannies and levels of the tidal flats. For example, sanderling rush about at high speed on the sandy flats and can pick up the tiniest shellfish and crustacea, including sandhoppers which are left on the very edge of the tide by each receding wave. Turnstones, with their short, stout bills and strong necks, can push aside stones and seaweed on rock and shingle shores to pick up the small shrimps, worms and sandhoppers hidden underneath.

Nearly all the invertebrates which live in the mud are filter-feeders, and need to sift the mud in water to obtain the tiny food particles within it – they all therefore have to live near the surface of the flats. Even the lugworms and ragworms which

179

An avocet searches the mud for food. Over 380 pairs now nest in Britain, thanks to a strong movement towards bird protection. The avocet is undoubtedly one of our most elegant waders, with a distinctive slender upcurved bill. Increasing numbers of avocets over-winter in Britain on estuaries in Devon, though they are normally summer visitors.

burrow more deeply must come to the surface every half an hour or so to cast sand. The only defence available to these animals against predators is burrowing and because they need to lie near the surface their chance of escape is limited.

The bills of the curlew and the godwit are long and sensitive and highly specialised, with a soft, sensitive tip full of nerve endings which can locate prey deep in the mud. They can probe 4in (10cm) down into the mud to find more deeply hidden prey. Curlews also use their long bills to turn seaweed in their search for small shore crabs, and to hunt for earthworms in nearby fields when the high tide has driven them inshore. Then there is the redshank with its bill of medium length, which has evolved two different feeding methods: it

probes the top 2–2½in (5–6cm) of mud for bivalves such as macoma and the amphipod, Corophium; and it will walk slowly over the mud delicately picking up surface-feeding worms and, in spring, the brown shrimp from high-tide pools. The shrimp is known to migrate southwards in winter so is only available 'in season'.

A familiar sight on most estuaries are the oyster-catchers, with their piebald plumage and long orange bills. Their name is quite inappropriate as they rarely eat oysters – their diet includes mussels, cockles and a related bivalve scorbicularia, as well as whelks and limpets. They use their bill to good effect, hammering holes in cockles and mussels to extract the contents. They have learned to penetrate the shells of bivalves and cut an internal muscle so that they can prise the shells apart. Unlike the bills of the curlew and godwit, the oystercatcher's long bill becomes harder as it reaches maturity so it can be used as a hammer.

Different species of wader feed at different stages of the tidal cycle. Thus the dunlin spends nearly 75 per cent of the time feeding when the mud is exposed, while curlews spend only some 47 per cent. Where birds have a nearly similar bill size, such as the knot and redshank, there are considerable differences in their activity cycle, the knot for example usually moving straight on to the flats while the redshank feeds first on the salt marsh; thus direct competition is avoided.

Some wildfowl eat the same food items as the waders; for example, shelduck eat the spire shells, hydrobia, as their main source of food along with other invertebrates found by dabbling in the surface layer. Pintail appear to be more estuarine than the other dabbling ducks, feeding extensively on the open mudflats on the ubiquitous spire shell, and also on the tiny bivalves, macoma, their slender bills being ideally suited for sifting out these creatures.

Mallard, teal and wigeon are all largely plant feeders, wigeon concentrating on Entermorpha algae and eel-grasses, while teal eat plant seeds, and mallard crop fresh marsh grass. Most wild geese feed on the grasses and stubble of the grazing marshes and farmland but Brent geese are primarily mudflat feeders, eating eel-grasses and green algae. However, the numbers of Brent geese have increased (due largely to protection from shooting) and they have recently taken to cropping grass and winter wheat in farm fields. Diving ducks can also be seen in estuaries in the winter, often in large numbers, the eider, scoter and long-tailed duck searching for mussels in sheltered bays, and the red-breasted merganser hunting fish on the high tide.

Birdwatchers out on the estuary foreshore in winter should be particularly careful not to disturb the birds particularly at high-tide roosts. Repeated disturbance means that the birds spend less time feeding, and if they are being frightened into lifting off time and again they quickly use up their energy, which will leave them even more vulnerable to cold weather.

Seals

The seal has great human appeal and is always a firm favourite with the crowd at the zoo, but far fewer people have seen one in the wild. By travelling to the right localities in Britain, however, it is relatively easy to see both the common and the grey seal, as, apart from the otter, seals are Britain's main coastal mammals.

The number of both seal species to be found in Britain is of international importance, since we play host to nearly 60 per cent of the world population of grey seals and 35 per cent of the European population of the common seal. Seals are very susceptible to human disturbance and prefer the quieter parts of the coastline. Physically, they are highly adapted animals designed to spend most of their lives in the sea but they are tied to the land during their breeding season, hauling out on their chosen beaches each year to give birth to their pups.

The last population estimate in 1984 suggested that then there were 85,000 grey seals around the British coastline, mainly concentrated in the north and west. Perhaps the best-known site is the Farne Islands, visited by many people in summer primarily for their large sea-bird colonies; however, large colonies also exist on certain Scottish islands such as North Rona where the exodus of people in the last hundred years has provided peace and quiet for the shy seal. Colonies exist down the west coast as far as the Scilly Isles, with sizeable populations on some of the Pembroke islands, such as Skomer. A very small number haul ashore on isolated beaches in Pembroke itself, and on the Scottish north coast, but as these cliff coves become increasingly frequented by leisure craft the increased disturbance may well drive them away.

The common seal, which is more numerous worldwide, but noticeably rarer in British waters, is mainly an east coast species; it is found from East Anglia north to Orkney and Shetland, seeming to prefer more sheltered waters and using estuaries, bays and sandy shores. Grey seals on the other hand are more often seen on rocky shores and islets. There are common seals in good numbers on the east coast of Scotland and especially in the northern isles, but the largest European colony

181

The grey seal is primarily a species of the rocky shores of the north, west, and north-east coasts. The grey seal is larger than the common seal and is more numerous in Britain where 60 per cent of the world grey seal population is found.

RIGHT
The common seal has its largest colony on the flat sands of the Wash in East Anglia. It has a shorter muzzle and a more domed head than the grey seal.

is found on the sandbanks of the Wash where as many as 6,500 have gathered in June and July. Common seals will sometimes travel up rivers, and single animals have been seen as far up the River Thames as Teddington. The common seal is also very shy and will usually disappear quickly into the water at the sound of a boat, although on the Wash they have become more accustomed to boatloads of tourists.

Identification is straightforward: the grey seal is larger and has a very distinctive straight-bridged 'Roman' nose, while the smaller common seal is rather more brown in colour and has a concave, slightly dog-like face. Both are expert swimmers, using their powerful back flippers to move at speed underwater. They can stay submerged for twenty minutes or more and their lungs and blood supply are particularly adapted to allow this.

Seals have been protected by law since the 1930s. However, on the Farne Islands in the 1960s the population tripled to nearly 7,000. The results were devastating: the mortality of the pups rose from 3 per cent to 25 per cent; the soft topsoil, the vegetation and the whole ecology of the islands were being badly damaged, and sea-bird colonies were being destroyed. Many hundreds of seals were culled, but even though this was done to protect the islands there was a popular outcry, especially as the Farnes have been a wildlife sanctuary since the time of St Cuthbert.

Grey seals breed in the autumn, the common seal in summer. The grey seal bulls arrive first and fights ensue to obtain the dominant position in the colony. The cows arrive a few days later – they are already pregnant, and soon give birth to a single white pup. This is suckled for three weeks onshore, and at this stage is unable to swim. However, it gains weight rapidly on the fat-rich milk of the mother and weighs some 5–6 stone (30–40kg) when it is finally left to fend for itself. The pups soon moult out of their white coats and leave the beaches. The cows will already have mated with the bulls again, and desert their pups; gestation is five months, but delayed implantation of the embryo occurs in seals to accommodate their rather particular life-cycle. The main grey seal colonies are on North Rona in the Outer Hebrides, Orkney and the Farnes; on average some 2,200 pups are born on North Rona, and 1,600 on the Farnes, but a smaller colony like Ramsay Island, Pembroke, raises only 200 pups each year. Research has shown that the life expectancy of the grey seal is some forty years for a cow but only twenty-five for a bull.

Common seals give birth to their pups on low-tide sandbanks and, unlike grey seal pups, these can swim – which they have to do at the following high tide, returning to the sands to suckle. There is no social or gregarious behaviour as with the grey

seal. It is possible to approach the pups on beaches quite closely, but the cow may stay on the beach to defend them and both can bite, the large and powerful adult quite savagely!

Both species feed mainly on inshore fish so there is minimal competition with the fishing industry. The effect of seals on fish stocks is tiny compared with the over-fishing practised by the industry itself, and the widespread pollution of the seas, especially the North Sea.

In the summer of 1988 a virus resembling canine distemper swept through the North Sea population of common seals, killing many thousands especially on the eastern coasts and in the very polluted Baltic. It quickly spread to the British colonies where periodic outbreaks have continued in ensuing years. Some scientists believe that the seals' immune system has been damaged by the high levels of industrial pollutants poured daily into the North Sea, leaving the animals open to infection; this has not, however been conclusively proven.

The Ecology of Sand Dunes

Sand dunes provide one of the most unstable and hostile habitats for plants and animals to live in, susceptible to wide variations in temperature and often suffering real drought conditions. Any rain falling on the upper surfaces quickly seeps through the sand particles, leaving growing plants without moisture and exposed to the heat of the sun. Despite this, sand dunes, especially the long-established ones, support a wide variety of life forms well adapted to the inhospitable conditions.

A sandy shore with offshore sandbanks exposed at low tide, together with an onshore wind, almost guarantees a coast fringed by sand dunes. They owe their existence to a group of tough plants, the most significant of which are the two stabilising grasses, marram and sea couch-grass. Sand dunes form when dry sand, blowing inshore, meets an obstruction. This is often the high-tide strand-line, occupied by flotsam, stranded seaweed and a thin belt of characteristic plants such as sea rocket, saltwort and orache. The embryo dune so formed on the lee side grows higher as more plants colonise and trap more and more sand.

These young dunes are called 'yellow dunes' because they are often of newly exposed sand, and they are highly mobile, moving inland at up to 23ft (7m) a year, even though to the casual observer they may seem 'fixed'. They take about fifty years to reach their maximum height and become 'fixed' dunes. Which plants colonise the growing dune is a reliable indication as to its stability – the further inland the more stable the dune, which results in a series of wave-like ridges and hollows that grow larger as they get farther from the shore, and are related to the plant types which colonise them. On the seaward edge of the young 'yellow' dunes only salt-resistant plants can grow, and here sea couch-grass is the most important pioneer, withstanding short immersion in seawater and binding the dunes with its roots. Next comes marram-grass which dominates the main dunes. It can grow a vast network of roots and underground stems, often growing upwards through several feet (a metre or more) of sand – this complex underground system effectively binds the dune together, making marram the most influential dune plant.

Once the marram is established other plants begin to grow on the more sheltered side of the dunes. All these dune plants have methods of preventing water loss. Marram leaves are dark green and rounded in dry weather, and grey-green and flat in rain; sea-purslane, saltwort and sea-sandwort all have fleshy leaves which can store water and which are most effective in shaking off the sand and salt.

The succession of plants to be seen is quite striking: first those on the new fore-dunes, then those on the marram-covered yellow dunes, and finally those on the older, flatter 'grey' dunes. Grey dunes are so-called because of the colour of the thick carpet of lichens which frequently covers large areas of them; they are also usually full of plants. To all intents they are 'fixed', unless they are adversely affected by forces which are beyond the norm such as fierce storms or excessive erosion – the latter includes the wear and tear caused by human feet.

Early fixed dunes are often still full of shell fragments blown in with the sand, and mixed with humus from decaying plants this results in a lime-rich soil. Such dunes therefore often support a superb variety of lime-loving plants more typical of chalk downland, as well as the plants adapted to living by the sea. Into the latter group come red fescue-grass, sand-sedge, sea-holly, sea-bindweed, sea-spurge and sea-beet; while the chalk flowers may include ragwort, lady's bedstraw, viper's bugloss, restharrow, yellow-wort and silverweed. Rarer plants such as grass of Parnassus may also occur.

Water will often accumulate in the older dune valleys, producing freshwater 'dune slacks' which, because they are already at sea-level, do not drain easily. Marsh conditions therefore soon occur here, and often become the richest part of the dune habitat with abundant marsh orchids, creeping willow and marsh marigold, and frogs, toads and newts in the pools. The rare natterjack toad is found in a few dune systems round the coastline, notably at Studland, in Dorset; Winterton, Norfolk; Ainsdale, Lancs; and on the Solway Firth.

Sea campion growing on a shingle ridge. Well adapted to the hostile shoreline environment, it is also found on cliffs. It has blue-grey fleshy leaves and many non-flowering shoots, which form a cushion. It flowers from June to August.

The surface temperature of the bare yellow dunes may reach a sweltering 150°F (65°C) on hot days, although at night the dew which is precipitated from the moist sea air is considerable, providing moisture for plants and animals. The plants support large numbers of insects and these also occur in reliable 'zones' since each different species is usually associated with a particular plant; and because dunes are often on coastal promontories, migrant butterflies such as the red admiral and painted lady are frequent visitors. Other herbivores also live on the plants – the banded snail, usually found on downland, is a frequent and brightly coloured inhabitant.

The main plant-eater on sand dunes is the rabbit, as large warrens are easily dug in the sand. Rabbits help to maintain the thick sward of plants and grasses so characteristic of the old-established dunes since their constant grazing always produces a short, dense turf. Field voles and wood mice live in the dunes too, and these attract the fox which emerges from the surrounding countryside at night to hunt them. And the hedgehog is a frequent visitor, especially if there are sea-bird colonies which it can raid. Large isolated dune systems provide particularly good sites for breeding sea-bird colonies: black-headed gull, common and Sandwich terns, and Arctic terns farther north.

The hedgehogs eat the eggs, and the foxes also prey on the chicks and the sitting adults.

Ringed plover and oystercatcher may breed on the dune edge, their eggs beautifully camouflaged against the mix of sand and lichens. The denser vegetation of the fixed dunes shelters skylarks, linnets and reed buntings; and in winter, flocks of finches, including twite and snow bunting from the north, will come in search of seeds, and also thrushes which feed on the orange berries of sea-buckthorn.

On the very oldest dunes, those farthest from the shore, the rain will eventually wash all the lime-rich shell fragments from the soil and after many decades it will be left nutrient deficient and slightly acid; as a result heathland conditions develop, with gorse, heather and bilberry eventually developing in succession to Scots pine and birch woodland. These dry seaside heathlands are a precious habitat and support some of Britain's rarest birds and reptiles; for example, Dartford warblers, smooth snakes and sand lizards at Arne and Studland in Dorset, and sand lizards at Ainsdale in Lancashire.

All dunes are subject to excessive wear and tear from the trampling feet of holidaymakers. The best are those with large bird colonies, such as at Blakeney Point in Norfolk, or with a superb flora, such as Braunton Burrows in Devon, and these are wardened nature reserves where the visitor is encouraged to respect this fragile dune environment.

The Ecology of Shingle Ridges

Shingle is one of the results of erosion and can occur from land surfaces anywhere; however, shingle beaches appear along exposed coasts where the sea has enough power and energy to throw up a pebble ridge. The products of erosion are mainly pebbles of different size and sand, and the simplest shingle beach is formed where the action of the waves sorts these out so that the pebbles are dumped at the top of the shore on the high tides, and the sand settles on the lower shore at slack water. These 'fringing shingle beaches' are common on parts of the south coast.

The second type is the shingle spit, where the prevailing winds and therefore the tides ebb and flow along the shore at an angle, this influence washing up the pebbles and eventually resulting in the formation of a spit which lies at a corresponding angle to the beach. This feature is known as 'longshore drift', and such spits may extend right across a bay in the form of a shingle bar, the best-known example being Chesil Beach in Dorset, some 20 miles (30km) long. Shingle spits may be quite complex in the way they have evolved down the centuries; sometimes successive ridges can be seen, often some way inland – these were thrown up a long time ago, probably the result of stormy conditions at sea. Such storm ridges are particularly noticeable at localities such as Dungeness, and Rye Harbour in Sussex.

Sand dunes suffer greatly from 'wind blow' which can destroy whole areas of old dunes; however, shingle ridges are much more stable. The greatest destructive element is man, digging for gravel and moving heavy machinery across the surface, all of which destroys the fragile flora. The creation of large gravel pits may lower the water table as water drains into them, and this will damage the plants growing at higher levels which depend on this water supply. On the positive side, these pits are frequently taken over by colonies of gulls and terns, as at Dungeness and Rye Harbour.

Although the surface of a shingle ridge is often dry, water – and especially dew – clings to the pebbles immediately below the surface. Soil particles are usually absent so the first colonisers are lichens and mosses, which can obtain their nutrients from rain and salt spray alone. These add to the humus collecting between the pebbles, and so other salt-tolerant species such as sea-sandwort may appear, and in northern Scotland the oysterplant may form a mat of vegetation. Once enough humus has gathered on the leeward side of the shingle ridge, other plants become established – the yellow-horned poppy, with its long curved seed-pods, and a waxy variety of curled dock with a tall brown fruiting spike. Eventually a carpet of flowers may cover the inner shingle ridge, with clumps of sea-campion, thrift and biting stone-crop. Here and there tall clumps of the uncommon

The yellow-horned poppy is found on many shingle ridges as it is tolerant of the salt-laden atmosphere. It flowers from June to September, producing an unusually long seed-pod which splits lengthwise when ripe. It is pollinated by bees and flies.

sea-kale may take hold, with gorse or bramble scrub on the oldest parts of the ridge, and sallow in the damper areas. On the shingle of north Norfolk dense thickets of the rare shrubby sea-blite, or *suaeda* have developed, a southern European species.

The birds which breed on shingle are specialists, and have to make use of the isolation and the camouflage offered by the stones. Ringed plover and oystercatchers are prime examples – their eggs and chicks are hardly visible even from a few feet (a metre or so), so well do they resemble their surroundings. On open shingle spits, free from disturbance, large colonies of terns may nest. Arctic and Sandwich terns may occur in vast numbers although little and roseate terns are much rarer. The black-headed gull will also colonise the shingle ridge in large numbers. This is a species which has increased dramatically in the last few decades because it has managed to adapt so well to man's environment.

All these species have well-camouflaged eggs and chicks, essential to minimise daytime predation from crows, large gulls and, in the northern islands, great and Arctic skuas. Stoats and foxes are the other most significant predators, taking sitting adults, eggs and chicks; these usually live in the surrounding countryside and make their way out to the shingle to hunt. Other mammals which are common are the rabbit and the hare, and these are resident, living amongst the vegetation of the open ridges.

Many sea-bird colonies are now carefully protected during the breeding season, roped off so that the casual visitor will neither trample the nests nor disturb the sitting birds; in some places, such as at Rye Harbour, electric fencing has been used to keep marauding foxes at bay. These measures have all helped to increase the fledging success of the colonies which, in the past, has often been very low. The little tern, in particular, was facing extinction in Britain, but has now recovered.

Bird Migration at the Coast

Bird migration is quite one of the most fascinating aspects of all zoology. Much is now known about migration due to modern scientific techniques, but some things still remain a mystery. How does the swallow, bearing its identifying ring, fly all the way to Africa and, the following spring, return to precisely the same farm shed to build its new nest? How is such a feat of pin-point navigation possible from such a tiny brain in such a tiny bird? No one can answer this at present.

It is now generally known from studies with high powered radar that birds migrate at considerable heights, often on a broad front and especially at night. Small warblers regularly fly at about 4 or 5,000 feet (1–2,000m) and larger birds at much greater altitudes. Birds navigate by the stars and moon at night, by the sun during the day, and also apparently use the earth's magnetic field. They learn by experience to make use of major geographical features such as coastlines, hills and river valleys which lie along their route; it is for this reason that bird observatories sited on exposed headlands, such as the one at Dungeness, are overflown by so many thousands of birds as they follow the line of the coast on their way out to sea.

Migration is a risk undertaken by a species to provide optimum conditions for the survival of the greatest number, which is why many of our birds migrate to Africa during the cold British winter. They arrive during the African rains when the availability of insect food is at its peak. Spring in Britain sees the return of the migrants, maximising their chances in the long days of our insect-filled, temperate summer. Similarly in late autumn birds arrive in Britain escaping from the rigours of the high Arctic and the extremes of northern European winters. Fieldfares and redwings from Scandinavia, blackbirds, song thrushes and starlings from Russia, geese from Greenland, Lapland and Iceland, waders from the whole northern tundra; all flock to these islands, still warm by comparison with the lands they have left behind.

Some birds, such as swallows, are total migrants,

the whole population changing areas with the seasons; others like the starling are partial migrants with a complex mixture of population movements. Millions of starlings move westwards out of eastern Europe and Russia in the autumn. The mean January temperature in London is 4°C (39°F); in Berlin it is –0.5°C (31°F) and in Moscow –10°C (14°F), so starlings migrate from the east because they would not be able to feed themselves during the snowbound winter months. Starling populations already living in milder western areas such as Britain have less need to migrate. The British winter starling population is larger by several million than during the summer, many birds spending nearly seven months here from October till April. It is thus a simple ecological truth that migrant birds are as natural a part of the bird fauna of an area as are any of the sedentary birds and some of these, like the swifts, may spend twice as long in their winter home in Africa as they do in Britain, where they breed between May and August, a period of less than four months.

There is great variation in the flight distance and direction of individual species but migration involves huge numbers of birds, with an estimate of 5 billion moving annually between Europe and Africa, plus the millions of winter migrants from the northern arctic areas and the transverse migrants leaving cold eastern Europe for the milder west. Some, like the warblers, may fly for two or three days and nights on end, while others like the swallows move in short daily stages, feeding as they move, and roosting at night. Many travel thousands of miles but the record-holder for long distances is the Arctic tern, which nests in places such as Shetland and Iceland in the north, and flies down to the edge of the southern icecap in Antarctica, maintaining a life forever in the sun, forever summer, by undertaking a round trip of some 25,000 miles (40,000km) each year of its life.

Some day-flying migrants fly at heights of below 200 metres and are therefore easily visible in many parts of the country. In London in the autumn of 1960 a team of observers kept early morning watch between mid-September and mid-November and counts suggested that over 4 million birds – starlings, larks, chaffinches, woodpigeons and many more – passed over the city arriving in Britain from the Continent.

Records made from visual observations are now supplemented by the practice of bird-ringing, in which a tiny numbered aluminium ring is placed around the bird's leg. The ring also carries a return address: in Britain, this is given as the Natural History Museum, London. This enables two points in the bird's life to be fixed; the place and date where it was originally marked and the locality and date at which it was subsequently found.

The recovery rate of many small ringed birds, such as swallows, is under one per cent, so ringing has to be regarded as a long-term study with results measured over many years. Recovery rates for birds like ducks and geese are much higher at between 10 and 30 per cent, due to the prevalence of game shooting in Europe; for the very conspicuous mute swan, the rate is 33 per cent.

Birds are caught at bird observatories by two basic methods. Most observatories use the Heligoland trap, a long funnel-shaped tube of wire netting over a line of bushes. The traps, usually six feet (2m) high with side wings at the entrance, taper to a 'catching box' after ten or fifteen yards (m).

More portable are mist nets, which originate from the Far East where they are used to collect wild birds for food. For bird-ringing purposes, these fine nylon nets are limited to about seven feet (2.2m) in height and made of horizontal panels. Placed in gaps in hedges and scrub, the nets catch birds flitting across the gaps. The birds fall into a loose pocket of netting, from which they are quickly extracted. They are then weighed, measured, sexed and sometimes even 'aged' according to plumage. Most birds are released within two or three minutes, when they quickly re-locate and return to their migration route.

Enthusiastic amateurs can learn to be bird ringers at bird observatories. The training period before a ringer is given his national licence to catch and ring birds is extensive; the activity is otherwise strictly illegal. The Ringing Office of the British Trust for Ornithology notifies recoveries to the individual ringer, perhaps a swallow from Cape Town or a starling from Moscow. Anyone can help with the 'recovery' side. Check dead birds, such as road casualties and tide-line corpses. Send the rings to the Natural History Museum, London, and you will be informed of the original place and date of ringing.

Insect Migration at the Coast

Between late spring and autumn many of the larger insects – the butterflies, moths and dragonflies – seen on Britain's south and east coasts may in fact be migrants. On fine, calm days in late spring waves of immigrant insects arrive from the Continent; and likewise in late autumn various insect species gather on Britain's coastline before migrating southwards.

These mass migrations of insects have been known about for centuries, and are even more noticeable in tropical countries. However, insects have been studied more closely in Britain for only some 150 years, and until recently it was thought that these insect movements were of fairly local dimension; it was erroneously believed that most insects live for only a day or two, whereas many in fact live for several weeks and some of the larger migrant butterflies, moths and dragonflies live for several months.

Insects are the most widespread and successful of all the animals, yet few people would associate the apparently aimless flutterings of a butterfly with migration over long distances. However, recent research has shown that certainly with some species this fluttering is not haphazard at all, and that regular migration forms an integral part of the life-cycle of emergence, feeding, reproduction and death of many temperate-region insects. Some of these, such as the painted lady butterfly, which move up from Africa and across Europe, are obviously more noticeable than others.

Any migratory movement of larger insects is most easily seen at coastal headlands, where they are concentrated by the landmass in the same way as bird migrants and often appear in large numbers at bird observatories like Dungeness. Many strong-flying insects are now known to travel all their adult lives, and some of the longest-lived, such as the red admiral and painted lady butterflies, may manage 600 miles (1,000km). This movement is known as a 'lifetime track' and usually occurs in one direction – a linear track. The insect, from its emergence, will fly in a more or less

The painted lady butterfly is an immigrant to Britain, originating in North Africa and spreading north across Europe each summer. The thistle is its principal foodplant, but the larvae also feed on burdock, viper's bugloss, mallow, stinging nettles and runner beans. The first immigrants arrive in May and June. They breed and produce more butterflies in July and August, as more immigrants continue to arrive from abroad.

constant direction depending upon the season, until on finding a suitable habitat it may stop to feed or even breed, sometimes for hours or days or sometimes for its whole life-span. If the terrain becomes unsuitable it will then depart in the same general direction by which it arrived, maintaining its overall lifetime track.

Many species, of course, rarely move from where they are born. Chalkhill blue butterflies in a stretch of downland may find all their requirements within one hundred yards; or meadow brown butterflies in a grass meadow may spend their brief four weeks or so of life commuting around the food-bearing flowers, all within a few square metres.

However, many do migrate so as to locate optimum living and breeding conditions in Europe. Experiments have shown that the peak direction of movement is north-north-west in spring, or towards the Pole, and south-south-east in autumn. This major change in the 'preferred direction' occurs at different times for different species. Of those species that regularly migrate over long distances the large white is the first to reverse its movement pattern; so the offspring of the eggs laid in May and June, which emerge as butterflies in mid- to late July, all show a tendency to the southward autumn movement. Small whites, red admirals and painted ladies all reverse their preferred direction in mid-August, while the clouded yellow butterfly is the last to do so, in early September.

The cause of this change of direction has also been established by experiment: with birds it is linked to the temperature and the lessening hours of daylight, whereas with insects the temperature and length of the night seems to be more important. Groups of small white butterflies subjected to an experimental 'long cold night' all headed south for the winter!

Butterflies and dragonflies migrate only by day, if the sun is shining; moths mostly migrate at night and may navigate by the stars, as do birds. Some moths, especially the well-known silver-Y moth, are day migrants. Silver-Ys may appear in thousands on the south coast in late summer.

As these migrating insects arrive on the south and east coasts they will move on inland, spreading out more thinly over the British Isles. In years when conditions are favourable – and the state of the weather in central and northern Europe is particularly pertinent – large numbers may reach Britain's shores. In leaner years they may be absent from much of Britain with only a sprinkling recorded in the southern counties. Although these migratory movements are most easily seen on the coast, a careful observer may be able to pick them up along prominent features which act as flight-guidelines, such as hills and valleys.

The most regular species of butterfly to be seen flitting in off the sea in a steady stream are red admirals, both large and small whites and painted ladies. Small tortoiseshells and peacocks also occur as migrants, and in most years a few clouded yellows appear. Some, like the admiral and the painted lady, originate from North Africa, starting their journeys in February. Rarer species such as Queen of Spain fritillary, long-tailed blue, and Camberwell beauty occur in small numbers most years; while the magnificent monarch is capable of crossing the Atlantic from America or the Azores if the winds are favourable. Hawk-moths are notable immigrants from Europe in late summer, the most frequent being the humming-bird hawk. Most others are rare – like the spurge, death's-head and oleander hawk-moths – although some ten species may occur. Dragonflies too may appear along coastal beaches, all heading in the same direction.

Migrations can be observed in even more spectacular fashion at some of the high mountain passes through the Alps and the Pyrenees, when millions of these self-same species of butterflies, dragonflies and even insects such as hoverflies may be seen hugging the ground up one side, flitting through the pass and travelling onwards to their unknown destination.

CHAPTER 8

SEA AND
ROCKY SHORES

LIFE on earth started in the sea and the sea still possesses the greatest abundance of living organisms; about 70–80 per cent of all plants and animals on earth live in the sea. These communities are rich and varied, especially those in coastal waters, and the natural forces which provide them with energy and nourish them are basically sunlight, and the ocean's own rich supply of soluble nutrients; the prevailing winds and the movement of water through currents and tidal flow are also of vital importance in the evolution of the ocean's life systems.

One of these currents, the Gulf Stream, originating in the Gulf of Mexico, is the greatest force affecting Britain's marine life. Its warming influence is largely responsible for the mild British climate.

The coastal seas lie over an extensive continental shelf which extends some 100 miles (160km) west of Britain itself; this means that all the inshore seas are relatively shallow, less than 650ft (200m) deep. Because these waters are so warm by comparison with the open oceans they are also rich in plant and animal life, and nearly all Britain's main fisheries are found in them. Indeed 90 per cent of the world's fishing is concentrated on a mere 5 per cent of the total ocean surface, although the oceans themselves cover 70 per cent of the earth.

Water temperature is the most important factor affecting the distribution of all marine plants and animals. Surface temperature readings show that the south-west coasts which face the incoming Gulf Stream are warmest, with a summer temperature of about 61°F (16°C), falling in winter to some 45–6°F (7–8°C). Despite its more northerly situation, north-west Scotland is still warmed by the North Atlantic Drift current and is in fact several degrees warmer all year than the south-east coast which is the coldest area. Mediterranean fish species are thus found in summer along the west coasts as far north as Cape Wrath.

In these shallower seas additional plant growth is promoted first by the sun's light and warmth, and also because of the higher concentrations of nutrient salts such as nitrates and phosphates which can dissolve in the warmer waters and are mixed by the inshore tides. Most of these plants are no more than tiny floating organisms, but they represent the first link in a food chain which stretches from animal plankton to large fish and sea mammals, and are a vital component in

LEFT
Lichens are zoned on a rocky shore; shown here are orange *Xanthoria* and the grey-green tufts of sea-ivory, *Ramalina*. Lichens are composed of two quite separate types of plant – an alga and a fungus. These grow together in a mutually beneficial partnership known as symbiosis.

the powering of the sea's life systems.

In a way similar to the terrestrial plants of garden and forest, marine plants create the various sugars, starches, fats and proteins needed to make living tissue by absorbing the necessary minerals, water and carbon dioxide through their external surface area, and by harnessing the energy of sunlight which they absorb through the green pigment, chlorophyll, in their leaves. This process also produces oxygen, which in turn supplies the needs of the marine animals. The great majority of marine plants are microscopic and drift with the currents. The most abundant are called diatoms and a single pint (½ litre) of seawater may contain millions.

The microscopic sea organisms which drift freely with the currents are all called 'plankton'; the plants are the 'phyto-plankton' and the thousands of tiny animals 'zoo-plankton'. The zoo-plankton feed on the phyto-plankton and many are also carnivorous, feeding on one another. Most are still microscopic in size. Some remain plankton all their lives, others are only temporarily so, usually larvae of crustacea, sea-urchins, jellyfish and many other large sea animals. As plankton they no more resemble their parents than caterpillars resemble butterflies. In winter, production of phyto-plankton – and consequently zoo-plankton – is at a minimum because the temperature of the sea water is at its lowest and the amount of sunlight is so limited. Both of these increase rapidly in spring and zoo-plankton especially are at their most numerous in the summer months. All the larger marine animals – from bass to basking shark, from grey seal to grey whale – are supported directly or indirectly by the plankton; so too are the millions of sea birds on surrounding shores.

Modern fishing methods may be very sophisticated, but in the past thirty years they have led to drastic over-fishing round the British coasts, especially of cod, herring and mackerel. Furthermore, pollution of the seas has become so bad that whole areas of the shallow Baltic and North Seas have become far less productive. The Baltic, in particular, has so much toxic waste that many regard it as the new 'dead' sea. Sewage from a huge and growing human population, oil spillages, and the more insidious pollutants such as polychlorinated biphenyls (PCBs), heavy metals such as cadmium and mercury and many more, are all dumped in the sea in ever-increasing amounts.

Rocky shores show more variation than any other type of marine habitat, from wave-cut platforms, steep cliffs, smooth slopes, cracks, crevices and pools to boulder-strewn beaches. What is most striking, though probably best appreciated from a distance, is the fact that plants and shore flora usually evolve in three clearly defined zones – so also, therefore, do the animals associated with them, although in a less definitive way. Each zone is identified according to the dominant species it supports; therefore the upper zone is called the 'black lichen zone', the middle is known as the 'fucoid' or 'barnacle zone' and the lower is called the 'laminaria zone' – this zone is below low-tide level and laminaria are large floating kelp seaweed. Only the middle layer is covered by every high tide and exposed by every low tide. The shore zones are thus closely related to tidal levels, and how the different species are distributed is relative to the amount of exposure each receives to both air and seawater.

Within each zone the plants and animals are greatly varied, but there is no one species that will tolerate the whole range of tidal extremes, from high to low water, of the spring tides. When the tide is high the temperature of the water is uniform and it provides a steady supply of oxygen, carbon dioxide and nutrients. However, as the tide recedes, plants and animals are exposed to much wider variations of air temperature and also risk drying out; thus their level of tolerance dictates exactly the level at which each will survive.

On exposed shores, wave action governs the composition of the plant and animal community; seaweeds do not become established to any great extent, and on coasts battered by waves, sea spray may extend the upper black lichen zone to 60ft (20m) above mean (average) high water. The upper zones will be dominated by the filter-feeding mussels and barnacles. On the other hand, in sheltered sea lochs with a small tidal range, the plants and animals of all three zones – lichens, barnacles, and mussels – may be compressed into some 10–12ft (3–4m).

On sheltered shores, great banks of seaweed may be seen growing and these, too, have evolved in zones, depending upon their ability to withstand drying out; they provide a valuable wet blanket on the shore when the tide is out, in their turn preventing animals from drying out, and also providing food for many herbivores. So, the first type of seaweed, usually found highest up the shore, is channelled wrack, well equipped to avoid desiccation because of its short growth and high oil content. Spiral wrack often grows with it, while lower down on the middle shore comes a band of knotted wrack whose long fronds are more able to withstand wave action. On more exposed shores bladder wrack, with greater tensile strength, replaces knotted wrack as the most common seaweed. Kelps, or laminaria, grow in the sea on the lower shore; they have pliable stems enabling them to survive in rough water, and some may grow several yards (metres) in length. Seaweeds are attached to the rocks and stones not by roots but by 'holdfasts', swollen basal discs fixed into cracks and crevices.

Below the brown seaweeds, but very common in deep rock pools, are the red seaweeds, with many more species and a variety of forms; some attach to the tougher kelps and wracks, and most are very delicate. Red seaweeds are the last plants to inhabit the depths before the fading light finally precludes plant growth altogether.

Animals of Rocky Shores

Inevitably, the inhabitants of rock pools and rocky shores lead lives dominated by the rise and fall of the tides; most of them rest at low tide, and feeding usually resumes with the return of the seawater – the life-system of filter-feeders like barnacles and mussels in particular is dependent upon the water which covers them at high tide. The wave action and its strength also have a noticeable effect upon the communities which occur on any particular shore. The lower part of a shore, below the tide line, is always more favourable to life as it does not suffer the extremes of exposure to air and seawater; competition to occupy it is therefore usually fierce, and some animals are often obliged to adapt to a higher zone.

Rock pools form natural aquaria in which animals can feed even at low tide. However, on sunny days the water in some pools may warm up considerably, and evaporation will result in a loss of oxygen as well as a rise in salinity; the converse is true if it rains. Because pools lower down the shore are subject to less temperature and salinity variation, they support a much richer variety of animals than higher pools. Anywhere offering protection from the buffeting of the waves on the lower shore, such as caves and large crevices, will almost certainly be full of rock-clinging animals, especially soft-bodied forms such as sea-anemones, sponges, hydroids and sea-squirts.

Some of the animals typical of the rocky shore may well resemble plants, living attached to the rocks and forming plant-like growths. Sponges, for example, are usually found as encrustations on wet rock, or sometimes hanging from crevices and overhangs; they are often brightly coloured. They belong to the very primitive group of animals called the Porifera, in which many similar animal cells adhere together in a set pattern, but there is no nervous co-ordination between them. They have a skeleton of fine spicules (needles) which can be seen with a lens and also felt if the sponge is probed gently with the fingers.

Sea-anemones and hydroids belong to the group of animals (called a *phylum*) called Coelenterates, where the body is integrated by a simple nervous system; they represent one stage higher in evolution than the sponges. Hydroids look very like colourful plants, the largest growing in sheltered crevices and cracks which protect their delicate structure from wave action and prevent them from drying out. They have a cluster of tiny tentacles armed with stinging cells around the mouth which catch minute planktonic food items. They tend to grow in colonies.

Sea anemones are related to the hydroids but are larger, and grow as separate individuals. They are radially symmetrical, and possess a large spread of tentacles equipped with stinging cells which paralyse their prey. They have to avoid drying out, so are found in pools and damp cracks and crevices. The commonest British species is the little red beadlet anemone, and the largest is the dahlia anemone which may be 3in (7cm) in diameter.

There are many segmented marine worms; some live under rocks and seaweed or in muddy gravel, and most are bristle worms, with tufts of bristles projecting from each segment. Some build tubes of chalky material, or of sand and shell fragments, and to catch their food they extend their fan of tentacles.

Starfish, brittle stars, feather stars, sea-cucumbers and sea-urchins are part of the large animal phylum called the Echinoderms. Most have spines – the name means 'spiny-skinned' – nearly all show five-rayed symmetry, and all have 'tube feet', tiny hydraulic organs with suckers on the end. Any with fewer than five arms are damaged, although

A spiny starfish in a rock pool. Starfish prey on other animals of the rock surface. The common starfish is widespread on the lower shore and can be variable in colour – some are a beautiful shade of violet. The brittle-star has longer, thinner arms which can fragment into pieces if touched.

new arms regenerate easily. Many are carnivorous and feed on a variety of molluscs and crustaceans. Sea-cucumbers are found mainly in the south-west and are soft and rubbery compared with the hard shell-like feel of the starfish but they do have the five sets of tentacles and tube feet of Echinoderms. They live in cracks and crevices and filter-feed.

The marine crustacea such as crabs, lobsters, shrimps and prawns are part of the largest phylum in the animal kingdom, the Arthropods, which number nearly a million known species and include all the insects and arachnids of dry land. They have a characteristic tough exo-skeleton, jointed limbs and a segmented body. Although a crab and a lobster look very different they are closely related, the crab being a flattened and shortened version of the lobster. There are some fifty species round British shores but only the common shore crab and the porcelain crabs are numerous. Crabs and lobsters are called decapods because they have ten legs; crabs have five rear pairs of limbs for walking and swimming and in both crabs and lobsters the first pair have developed as claws. These are used to collect food by scavenging.

A shore crab, camouflaged by seaweed, in a rock pool. There are several dozen species of crab around Britain's coasts, the most common of which is the shore crab. Another common resident of our shores is the hermit crab – in itself a defenceless creature since its body is unprotected by the usual carapace. However, it protects itself by occupying the shells of winkles, whelks and other marine molluscs.

Barnacles are unique amongst crustacea, their limbs modified to form a web of feelers which pick up food from the currents; they are attached to the rock by their heads. Their larvae are tiny plankton which, when settled on a rock, go through a metamorphosis and develop a shell. Acorn barnacles are often the most numerous things to be seen on an inter-tidal rocky shore and can number up to 30,000 per square yard (metre).

Nearly all the animals on the shore with shells – including winkles, whelks, chitons, bivalves and tusk shells – form part of the animal phylum Mollusca, or molluscs, another huge group with over 100,000 known species. Land molluscs include the snail and the slug. Features common to both sea and land molluscs are that most secrete a calcare-

198

ous shell from a mantle of muscle which covers the important parts of the body like a tent, and that they move using a flat sticky pad or foot, although the squid and octopus have evolved a stage further and use water jet propulsion.

One of the most common shore molluscs is the limpet, and there may often be as many as three species of limpet present on the same piece of shoreline. On still days it is possible to hear them feeding, as they scrape algae from the rocks with their rasping tongue. Other abundant and conspicuous molluscs on rocky shores include periwinkles, topshells and dog whelks; they are all characteristic of this habitat and occur in distinct zones, especially the periwinkles and topshells, with different species at different levels on the shore. Bivalves are more typically the molluscs of sand and mud, but mussels are very successful on rocky shores, forming great beds. Other bivalves are rarer; scallops, cockles and Venus shells may be present, but bivalves do best on softer shores.

Small fish are often numerous in large rock pools, and are superbly camouflaged to merge with their surroundings. Most are only 2–3in (5–8cm) long, and have evolved with thin, elongated bodies to hide in cracks or under stones. Butterfish, blennies, gobies and the fifteen-spined stickleback are the most common.

Insects are rare on the shore, but there are two which are noteworthy: the marine bristletail which lives in splash-zone crevices, and the slate-blue springtail which survives twice daily tidal submersion, reappearing to crawl over the rocky surface.

Plants of Rocky Shores

Cliff plants often produce a spectacular natural rock garden, but how successful they are is dictated firstly by their situation in relation to the prevailing wind, then by the amount of salt spray they receive, by the geology and slope of the rocks, and finally by grazing animals, especially rabbits and sheep. The flat expanse of a salt marsh could hardly be more different from the nooks and crannies of a steep cliff, yet the sea and the salt-laden air provide a link and some plants, such as thrift and buck's-horn plantain, are found in both.

For flower species confined to the coast the main problem is the supply of freshwater, and this applies both to salt-marsh and cliff plants. Cliffs especially are subject to strong and drying winds which increase the rate of water loss of the plants; many have evolved ways to prevent this loss; their leaves are thick and leathery or with a waxy or hairy surface, or they have a reduced surface area, or they grow in a tight basal rosette. Furthermore, the prevailing winds on the coast blow with unremitting constancy, and this is a considerable influence in suppressing plant cover – you need only examine both sides of a rocky headland. Plants adapted to saline conditions are known as halophytes and most plants found in the splash zone of a rocky or cliff shore are of this type, able to survive where the amount of seawater in the soil may be as high as 30 per cent.

Cliffs which are easily weathered support a much richer flora than those made of hard, granite rocks. The maritime lichens which grow on their steep rock faces form valuable humus when they die, giving nourishment and further foothold to flowering plants in crevices and on ledges where soil accumulates. From the shore, on the rocks immediately above the level of the seaweeds, are salt-tolerant lichens, often in distinctly coloured bands which are an indication as to the level of inundation by salt spray. The first layer is of a black encrusting lichen, Verrucaria, which may be mistaken for an oil stain. Above this is a band of the bright orange lichen Xanthoria, which merges into a grey encrusting band of Ochrolechia, often with grey-green tufts of sea-ivory, Ramalina.

The most widespread of all maritime plants is thrift, or sea-pink. It is found all round the British coast, on sea-cliffs of all types, and on salt marshes, as well as inland on mountains. It will sometimes grow in rock crevices among the upper lichen bands, but more often above them.

Unlike salt-marsh, dunes or shingle banks, cliffs have no clearly defined community of plants, although several species are characteristic. As well as the ubiquitous thrift, these include golden samphire, sea-campion, buck's-horn plantain, sea-beet, common scurvy grass, rock sea-lavender and

199

rock samphire, the latter being replaced in Scotland by lovage, a related umbellifer. The commonest grass is often red fescue, also found in other coastal localities because of its salt tolerance.

Notable additions to this flora occur on the west coast from Devon to the north of Scotland, influenced as it is by the warm, moist Atlantic climate. Wild cabbage is a notable western plant, the ancestor of all the modern cabbages of the farm. In spring the short turf of many western cliff tops, especially on undisturbed islands, is filled with a carpet of blue spring squill and the closely related common bluebell. In sheltered parts of the Atlantic coast the striking tree mallow, up to 10ft (3m) tall, may also occur, while in cracks and crevices rock sea-spurry, English stonecrop and the fern sea-spleenwort may appear.

Low down on the cliff, salt spray is the most significant influence and severely limits the plant species which become established, but higher up out of reach of the spray several plants may occur which are typically found in walls or stone surfaces, such as wall pennywort and ivy. On acidic rocks a sub-maritime heath-plant community may develop, such as on the Lizard in Cornwall where Cornish heath grows in profusion on the rock type there known as serpentine. Basic rocks such as limestone will support a much wider variety of plants, creating a colourful rock garden: sea-carrot, biting stonecrop, thyme, kidney-vetch and both red and sea-campions. The cliff tops may be a valuable wildflower reserve.

In some parts of the far north and west of Scotland the cool, moist climate enables a number of arctic-alpine plants to grow at low altitude on sea-cliffs. Several saxifrages, moss campion, roseroot and even mountain avens can all be found nearly to sea-level in favoured localities.

On cliffs and headlands occupied by sea-bird colonies the sparse soil will be vastly enriched by the nitrogen and phosphorus from the birds' droppings. However, while a light extra 'manuring' of the soil may dramatically increase the number of species able to gain a foothold, too much is detrimental, and a heavy inundation of bird guano will favour only a few vigorous species. The plant community is then dominated by scurvygrass, sea-beet, oraches, stinging nettles, goosegrass, chickweeds and sorrels, all of which cope well with the increased nitrate levels.

Sea-birds

In 1969/70 a first-ever attempt was made to count the sea-birds breeding in Britain. Called 'Operation Seafarer', the project showed that although some species were obviously thriving, the totals for others were actually much lower than previous haphazard estimates had led one to believe. And since that time, although a number of species have continued to increase and expand their range (primarily the fulmar, kittiwake and gannet) many are still declining, especially the auks and, notably, the puffin. Terns, too, have in general lost ground. Puffins are, of course, difficult to count because they nest in burrows, but singletons on cliffs usually indicate a pair in residence. At the turn of the century their population was estimated at 5–10 million pairs; there are now only 500,000 pairs in the British Isles and numbers are still declining. Counts for the three nocturnal burrow-nesting species – Manx shearwater, and storm and Leach's petrel – were all rough estimates only. The most common sea-bird was the guillemot, with 577,000 pairs; followed by the puffin at 490,000, the kittiwake at 470,000 pairs, and gannets at 138,000 pairs. The rarest were the Arctic skua at 1,100 pairs and the little tern at 1,850 pairs. Since 1969 the gannet has increased to 160,000 pairs, and the little tern having sunk to around 1,000 pairs, is also beginning to recover thanks to successful protective measures.

In 1988 these populations suffered a serious setback as many northern colonies failed to produce chicks because there were no sand-eels to feed them on. One again, man was responsible as sand-

Pink thrift is a common plant found on all types of shore, from shingle ridges to steep cliffs. Also known as the sea pink, it flowers from April to August, its colour varying from white through pink to red. Great drifts of flowering sea pink make a glorious sight among short wiry cliff-top grass, and they are hardy colonisers of inhospitable rocky sites.

eels had recently been mass-harvested for fish meal in the North Sea, the stocks of the large commercial fish having run low through over-fishing. It is quite possible that pollution of the North Sea was also a contributory factor. There is also concern that sea-bird populations generally are increasingly threatened by the growing number of oil-fields that Britain has round its shores, and by the insidious and potentially disastrous problem of sea pollution by chemicals.

The twenty-four species of Britain's sea-birds divide into six separate groups – six species of gull, five terns, four auks, four petrels, two skuas, two cormorants and the gannet.

The most familiar sea-bird is the black-headed gull, especially during the non-breeding months from August till March when it may be seen even in the heart of the city. At this time it does not have a black head but only a winter-plumage black spot behind the eye. Despite its familiarity it lies only eighth in total breeding population, although many thousands arrive from the Continent in autumn to winter here; counts in the London area alone showed 250,000 roosting at the large reservoirs around the city. Their breeding colonies are scattered throughout the country, although there are hardly any in Wales and the South West. Large colonies exist at Needs Oar Point in Hampshire, along the East Anglian coast, and in Orkney, with a total British population of 200,000 pairs. They eat invertebrates, especially on farmland where they take large numbers of leatherjackets, wireworms and the like. Winter totals now exceed 1.8 million birds.

The herring gull is probably the next most common species, and like the black-headed gull it has fully exploited its association with man, nesting on the rooftops of many seaside towns as well as in large colonies on isolated shingle banks or broad

Lesser black-backed gulls, here forming part of a huge mixed colony of 40,000 pairs together with herring gulls at Walney Island, Cumbria. The lesser black-backed gull can be distinguished from the great black-backed gull by its yellow rather than flesh-coloured legs, and its dark grey rather than black back, although colouring does occasionally vary. The principal difference, however, is in size; the lesser black-backed gull is about 21in (53cm) while the great black-backed gull is 25–7in (63–8cm).

cliff ledges. The herring gull population is some 200,000 pairs. Large colonies exist round all the coasts except in the East Anglia area, and these colonies may totally change the local ecology, driving out many other species. These gulls will feed on almost anything – earthworms, grain, turnips, birds' eggs and even their neighbours' own chicks.

Lesser and greater black-backed gulls are distributed widely in small numbers, with one or two particularly large colonies. In Cumbria 22,000 lesser black-backs share Walney Island with herring gulls, while the largest colony of great black-backs is some 2,000 pairs in Orkney. Both species are relatively large, and because of their size frequently take the eggs, young and even the adults of many other sea-birds; they also scavenge on fish waste and offal. Great black-backs tend to be more solitary nesters, single pairs occupying their own little offshore stack (rock) or islet. Some 50,000 pairs of lesser and 22,000 pairs of great black-backs were counted by Operation Seafarer.

Despite their name, common gulls are in fact the second rarest of the six species with only some 47,000 pairs, nearly all breeding in the far northern and western parts of Scotland. The largest colony is in Aberdeenshire where 3,000 pairs are found on a large grouse moor. Far more winter here, with some 700,000 having been estimated in 1983.

The most numerous gull, and third in the sea-bird league table with some 490,000 pairs, is the elegant little kittiwake, the most oceanic of British gulls. Huge colonies nest on sheer cliffs, all the way up the east coast from Dover in Kent to Unst in Shetland. In the absence of sheer cliffs in East Anglia, they have colonised ledges on buildings on the Lowestoft sea-front. Surveys in 1959 and 1969 showed an annual increase of 4 per cent, probably now much reduced. Their prime food during the breeding season is small fish, with sand-eels most common. By late autumn most of the population has returned to the waters of the North Atlantic.

The auks, previously estimated at millions on St Kilda alone, are the most numerous British sea-birds, although both puffin and guillemot only reach half a million pairs each in total, nearly all in huge cliff colonies around the coasts north and west from Yorkshire. These species, along with 100,000 pairs of razorbills, occupy their breeding ledges early, with guillemots first back by New

Year, razorbills by February and puffins *en masse* in late March or early April. The black guillemot is much rarer; it is almost exclusively a Scottish bird, and lives in smaller scattered colonies on low cliffs – only some 8,000 pairs were estimated in 1969. All the auks feed on small fish such as sand-eels and sprats by diving from the surface of the sea and propelling themselves underwater with their wings.

Cormorants and shags also feed by diving from the surface, but they swim underwater using their large feet as flippers. There are some 36,000 pairs of shags, which nest on rocky coasts, mainly in Scotland; the 6,200 pairs of cormorants are more widely scattered. Both build untidy nests of sea-weed and sticks on wide ledges. They feed on fish including sand-eels and flatfish, especially dabs, and remain in coastal waters throughout the year.

Five species of tern breed in Britain, but they winter in more southerly climes. Roseate and little terns are rare with 2,400 pairs of little terns and, by 1990, only 330 pairs of roseate. Roseate terns are restricted to small islands, usually in sheltered waters such as the Firths of Forth and Clyde, while little terns breed on shingle banks and are very susceptible to human disturbance. Common and Sandwich terns are scattered in many coastal counties but are absent from the South West except for the Isles of Scilly. The largest colonies are on the east coast, on flat sand and shingle banks, and some 12–14,000 pairs of each species occur. All feed on small fish with some Crustacea and marine worms, and they catch the fish by plunge-diving from some 15–35ft (5–10m) above the surface. The Arctic tern is a more northerly species and numbers some 80,000 pairs, usually found in the northern islands; here, however, they are preyed upon by both great and Arctic skuas. The great skua is largely limited to Orkney and Shetland, while Arctic skuas also breed in the Outer Hebrides; both are piratical sea-birds, chasing others and forcing them to disgorge their catch. They will also feed on adults, especially puffins which are particularly vulnerable at their cliff-top burrows. Skuas nest among the heather on the island moorlands; the great skuas's range has expanded considerably, although the Arctic appears to be in decline.

The Manx shearwater, storm petrel and Leach's petrel are all oceanic birds which in winter travel far down into the southern and western Atlantic. In spring they breed in isolated colonies on islands from Pembroke north to Shetland. Because they nest in holes and are nocturnal their populations have proved difficult to estimate, but Manx shearwaters certainly occur in large colonies with some 50,000 pairs on Skomer and 100,000 on Rhum.

The gannet and in particular the fulmar are both success stories, with rapidly expanding populations. Until 1878 the fulmar was almost non-existent in Britain, except on St Kilda; now, some 520,000 pairs occur round Britain's coast, although the largest populations are on the northern cliffs. In Shetland they even nest on low stone walls. The gannet with its 6ft (2m) wing span and distinctive cigar-shaped body is Britain's largest breeding sea-bird. Its population has trebled in British waters since the early 1900s and Britain now plays host to three-quarters of the world population. Gannets are magnificent birds which breed in squabbling noisy colonies mostly on islands, with the sole exception of 650 pairs at Bempton in Yorkshire. Like terns, they plunge-dive for fish from up to 100ft (30m), feeding their single chick for three months till fledging; the first-year birds migrate to winter off the coast of West Africa.

Sea-bird Colonies

Although the oceans cover some seven-tenths of the earth's surface only about 300 of the 8,600 bird species – ie about 3.5 per cent – are sea-birds. Clearly this is because conditions on land are far more diverse than those at sea. Some 95 per cent of these sea-birds do breed in colonies, but since most of them are primarily adapted for life at sea they are awkward on land and this makes them vulnerable to terrestrial predators, from hedgehogs

Fulmars have expanded their range round all of Britain's coasts; in Shetland they even nest on old stone walls. Although in appearance they are rather gull-like, they are actually members of the petrel family. In flight they are a joy to watch, as they soar close to the cliff-edges and skim the sea.

and rats to man himself. It is for this reason that they very often choose to establish their colonies on offshore islands, isolated headlands or sandbars.

Sea-birds avoid competition for nest sites by showing distinct preferences for certain niches. They are not normally influenced by the vegetation of a cliff, spectacular though this may be, but depend more on its physical nature.

Puffins prefer the cliff tops because they like to nest in burrows, and the deeper, softer soil on the cliff-top surface enables them to do so. Sometimes they use old rabbit burrows, but they are very capable of digging their own. Where colonies are dense, care should be taken when walking the cliff edge to avoid collapsing the burrows which riddle the peaty soil. Manx shearwaters and storm petrels also use such burrows but often nest among stones, such as the scree slopes of hillsides on the island of Rhum.

The other auks nest lower down the cliff face but still avoid each other. Guillemots crowd together in thousands on open, exposed and often very narrow ledges below the lip of the cliff; razorbills avoid open ledges, living in cavities and crevices, while black guillemots nest on low cliffs in holes or, more often, under boulders on storm beaches. Both guillemots and razorbills require cliffs which fall vertically into the sea, as their chicks leave the nest before they are fully fledged by jumping straight into the water.

The gannets will often make their pitch just below the puffins, nesting in large, evenly-spaced colonies on broad cliff ledges, steep rocky slopes and sea stacks. They build nest-mounds of seaweed and flotsam and perform dramatic displays of greeting and courtship at the nest site.

Colonial breeding has some clear disadvantages: a disturbance which causes general panic, for instance, may result in the loss of many eggs and chicks; and predators can obviously detect a large colony easily, and tend to concentrate their own populations around it. Nevertheless, colonial breeding does have its advantages, perhaps the most significant being firstly, the fact that many watchful eyes may apprehend an intruder more readily, and many birds may harass it more effectively than just one or two; and secondly, the synchronised laying of eggs, which cuts the loss rate of

eggs and chicks from predators since they can only eat a finite amount at one time. It also means that any rise in the available food supply will benefit the whole colony, not just one or two individuals; and it reduces potentially harmful competition between adult birds for the simple reason that they are all too busy with their young! It is also thought that colonial breeding provides 'social advantage', and that all the courtship activity and display stimulates hormone production and other preparations for breeding so that reproductive success is improved.

Once the chicks have fledged their mortality is high, the first year being by far the worst when as many as 70 per cent may die. Once immature sea-birds have survived beyond their first migrations and have entered their second year, their life-expectancy improves although only 5 per cent may survive to adulthood. Many sea-birds take several years to reach maturity: terns may breed in their second year, but kittiwakes only in their third or fourth, and fulmars not until their seventh or eighth years. However, their life span is quite considerable, which compensates for their long period of immaturity and low reproductive rates, with gulls and gannets reaching twenty-five to thirty years of age, and fulmars possibly longer.

Kittiwakes use tiny ledges and rocky outcrops, plastering their nest platform to the tiniest obstruction, and forming huge colonies. Their population has expanded dramatically and now they even nest on buildings, for example in South Shields or on the pier at Lowestoft. Fulmars are also managing to extend their numbers, and to some extent their nest sites overlap with those of the kittiwake; however, they seem to prefer wider grassy ledges and rarely nest among the auks, forming well-spaced-out groups rather than large closeknit colonies. Both the sitting adults and the young fulmar chicks have a particularly individual defence against predators, highly effective against

Puffins prefer to nest on the cliff-tops, digging burrows in the soft soil or taking over disused rabbit burrows. The puffin's colourful parrot-like bill, magnificently marked in red, yellow and blue in the summer, is partly shed in winter, and is capable of holding several sand-eels at a time.

a lunging peregrine, and certainly capable of distracting a human – they spit oil, an unpleasant, fishy, sticky oil which clogs and clings.

The location of nesting sites for both the shag and the cormorant is dictated by the fact that unlike other sea-birds, their plumage is not waterproofed by oil and so it becomes wet through from constant immersion when fishing. They therefore tend to choose their site down near the base of the cliffs, the shag usually among fallen rock debris and on the lower ledges where it can easily return when its wings are sodden. Cormorants are more likely to nest apart from these huge colonies, but they too usually occupy low stacks and rocks. Lines of shags and cormorants may be seen sitting out on the rocks drying their wings.

Large gulls, especially great black-backed gulls, always make up part of these sea-bird colonies and are major predators of eggs and chicks. They are much more adaptable in their breeding sites, although great black-backs tend to occupy solitary stacks and islets.

The life cycle of most sea-birds is divided into two parts, the five months from late February or early March until late July, during which they are tied to their breeding colonies, and the other seven months when they are far out at sea. As they leave their breeding colonies, large numbers may be seen passing many of our headlands on their long migration out to the open Atlantic.

Sea-cliffs also hold land-birds such as ravens and jackdaws which forage inland for their food; and in the far north, wild rock doves. A true sea-cliff nester is the chough, but it is now rare and restricted to a few places in Wales, Ireland, the Hebrides and the Isle of Man, its total population only about 800 pairs.

On the highest and most remote sea-cliffs a few pairs of golden eagles may be found, and perhaps even the reintroduced sea-eagle, of which about 70 have been released; also the peregrine, all these

Guillemots are the most numerous of Britain's seabirds, with a population of about 500,000 pairs. They swim on the sea – often in large rafts – and dive to catch fish below the surface.

Guillemots breed in large colonies on cliff ledges and eggs are laid directly on to the bare rock; the eggs are pointed at one end so that they roll in a circle if disturbed.

raptors preying on the other sea-birds, taking full advantage of such an easily available food supply. And there may be another resident high on the cliffs, as much a part of the cliff colony as the puffin, a little brown bird with a lusty voice singing from a rock amidst all the hustle and bustle of such vast sea-bird numbers: the ubiquitous wren.

ABOVE
Gannets nest in large colonies on isolated islands. The only mainland colony is at Bempton Cliffs, Yorkshire. When fishing, gannets dive from 100ft (30m) or more, sending up showers of spray as they enter the water. They can often be seen methodically diving one after another when a shoal of fish has been located.

RIGHT
Kittiwakes bathing in a freshwater stream at Fowlesheugh, Grampian. This elegant little bird is the most numerous and oceanic of British gulls, nesting in huge colonies on sheer cliffs. Undaunted by the absence of cliffs in East Anglia, they have colonised ledges on seafront buildings, and now nest on the cliffs of Dover.

Sea-fish

Several hundred species of fish live in the seas around Britain and some fifty or so varieties are caught commercially. In the early days of his evolution, man lived as a hunter-gatherer; today, commercial fishing is the sole surviving method of food production involving true hunting.

The warming effect of the Gulf Stream waters on what would be cool sub-arctic seas means that Britain's sea-fish come from two sources; many species are arctic and northern temperate fish, while a number are Mediterranean and sub-tropical fish which migrate northward into British waters in the summer. Among these are several sharks, the tunny and the pilchard.

Most sea-fish lay eggs which float, and when they hatch the larvae and the eggs form part of the temporary plankton in the surface layers of the sea. Here the tiny fish fry eat microscopic zoo- and phyto-plankton, and are themselves eaten by many larger fish.

Commercially the two most important groups are fish of the cod family and the flat-fish. Some twenty species of the cod family are found regularly in British waters including the cod itself, the haddock and the whiting. The cod, the most important British fish worth some £60 million per year, is an arctic species extending throughout the Arctic, North Atlantic and down to the Bay of Biscay. It spawns in spring, its fry living in the upper layers of the sea until they are some three months old. They then go deep and tend to mature at depths of up to 2,000ft (600m), well below the tidal zone. The total North Atlantic catch from all countries is estimated at some 2½ million tons per year. The haddock has a similar distribution but spawns at greater depth. While the cod eats other smaller fish as well as bottom-living crustacea, haddock are known to feed more on worms, starfish and molluscs. Whiting, pollack, ling and pout are all landed in commercial quantities; they tend to be more inshore fish than their larger cousins in the cod family.

Flat-fish include such important species as plaice, sole, turbot, flounder and halibut; these are bottom-living fish usually found in shallow water and are most numerous over a sandy sea-bed. The young larvae live among the plankton in the upper sea layer like other fish, and also swim upright, but after a few weeks they undergo a major biological change as one eye 'migrates' across to the other side of the head. The upright young fish then turns on to its side and grows into the full adult form. It spends much of its day part buried in the sand and mud of the sea-bed, protected by its upper coloration which camouflages it perfectly. The other flat-fish, the skates and rays, are flattened from the top, so to speak, with eyes set normally on top of the snout. Rays are part of a large group of fish known as cartilaginious fish, one which includes the sharks and dogfish, distinguished because they have skeletons of cartilage instead of bone (as in the bony fishes).

Several species of shark visit west-coast waters, including the world's second largest fish species, the huge basking shark which may grow to 40ft (12m) in length. Despite its size it is quite harmless, filter-feeding on plankton which it sieves through its gills. Basking sharks usually appear off west-coast beaches in summer, often cruising close inshore. The tunny is another large southern fish which migrates north in spring, feeding on mackerel, herring and other fish.

Many of the commercial fish species have been disastrously over-fished, and the herring is the classic example. It used to be very numerous around the east and north coasts of Britain, but once it was learned that there were several races of herring which spawned in different places at different times, they were pursued relentlessly by modern trawler fleets until the annual North Sea catch had declined from around 1 million tons in the late 1960s to less than 50,000 tons in 1977. In the past, spawning shoals might have been a couple of miles (several kilometres) long, containing 2–500 million fish. The herring is the only commercial fish to lay eggs which sink to the bottom. On hatching, the larvae move to inshore and estuarine waters along with various other important young fish species, moving slowly back to sea over the following five to six years as they mature.

The mackerel is a southern species which migrates as far as the north coast of Scotland in summer. There is also a population in the North Sea, though little is known about the migrations of these. Since the decline of the herring, the fishing

Plaice are flat-fish which spend most of their lives lying camouflaged on the sea bed. They begin life with one eye on either side of their body in normal orientation. Then, in a remarkable biological change one eye migrates to the other side of the head so that both are on top where they stay for the rest of the fish's adult life.

industry has turned its attention to the mackerel, and fishing fleets from many countries congregate off Britain's west coast. The stocks of mackerel could very well therefore go the way of the herring.

The body shape of a fish is closely linked to its life-style. Bottom-dwellers are usually flattened with their top surface well camouflaged. Mid-water fish like the mackerel and herring are built for speed, with muscular, streamlined bodies. Most mid-water fish live in shoals, both as a defence against predators and for increased efficiency when 'sweeping' plankton for feeding.

For most people their only glimpses of live marine fish occurs when they are on holiday and paddling in rock pools, when they may see tiny fish darting through the seaweeds to hide under stones and in crevices. These are the colourful but superbly camouflaged blennies and gobies, well adapted by size and shape to their sea-shore life-style. Blennies, gobies and similar shore fish are of no commercial value, but they are important to the coastal birds, and to seals and otters. Sand-eels are also a particularly important food for many of Britain's breeding sea-birds, and used to occur in large shoals around the northern coasts. The eels are quite small, only 3–4in (8–10cm) long, but they are now being fished in great quantity for fish-meal products because other species have been so extensively over-fished.

Whales

Whales range from the huge blue whale, at 100 tons the largest animal which has ever existed, down to the common porpoise, smaller than a man and only a few stones in weight. However, they have all developed many specialised features to deal with both the advantages and disadvantages of living permanently in the sea.

Although whales look like large fish they are true mammals, warm blooded, breathing air through lungs and giving birth to live young which suck milk from the mammary glands of the mother. Another distinction is that their tails beat up and down, rather than sideways as in fish. There are about eighty species of whale, all belonging to the zoological order Cetacea; 86 per cent belong to the sub-order Odontoceti, the toothed whales, the rest belong to the Mysticeti or baleen whales. The most obvious difference between the two groups is in their feeding systems: teeth in the Odontoceti, and whalebone or baleen plates designed to filter-feed for plankton and small fish in the Mysticeti. The toothed whales have an asymmetrical skull with only one 'blow hole' – the blow hole being the whale's nose – and the baleen whales have two symmetrical blow holes, or nostrils. Most whales live in herds known as 'schools', and are very gregarious and family orientated.

Both porpoises and dolphins are toothed whales, Odontoceti, and so is one of the largest whales, the sperm whale which may reach 60ft (18m). The killer whale and pilot whale also belong to this group and are large members of the dolphin family.

Most of the baleen whales are very large in size, and fall into three major groups – the rorquals, which include the blue and the humpback whale, the right whales and the grey whale. This great size may be an advantage in storing fat reserves and withstanding predators, but it is only possible because they live in the sea and their enormous

The bottle-nosed dolphin is one of the most frequent visitors to British coasts among the small whales. Dolphins are distinguished from porpoises by their pointed beaks and sharply pointed back fins. They are usually seen when they leap energetically from the water.

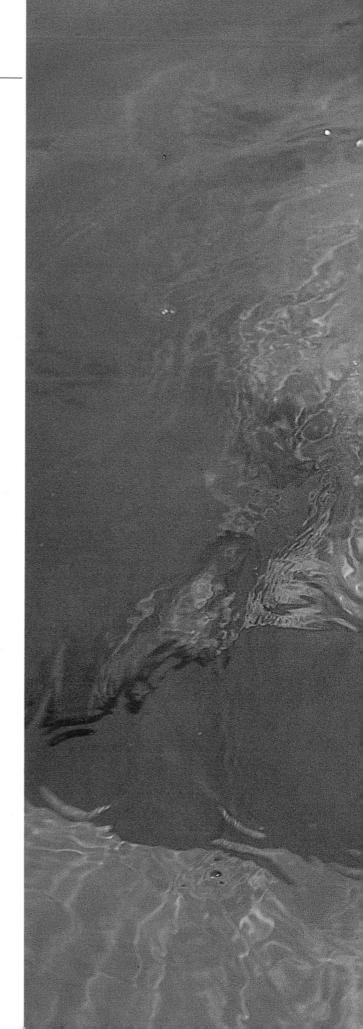

bodies are supported by the buoyancy of the salt water.

There is evidence to show that whales evolved from land mammals, but if they were to take to terra firma now they would need huge limbs to support their weight. Large whales stranded ashore die rapidly because their ribs collapse, unable to support the body weight out of water. They have evolved a very mobile, streamlined shape with a horizontal boneless tail fluke and two horizontal flippers. Most have an upright dorsal fin to assist with stabilisation. Forward power comes from the tail which beats up and down, most forward propulsion coming from the upstroke. The flippers have a skeletal structure similar to the human arm and are used for steering. Whales have no hind limbs, although vestigial traces may be found in the bony skeleton inside the body. Instead of the land mammals' hair to keep them warm, they possess a thick layer of fat or blubber, which may be a foot (30cm) thick in large whales. On the other hand, they have no means of cooling themselves and small whales like dolphins may die from heatstroke should they become stranded.

Whales have to return to the surface to breathe but some, like the sperm whale, can dive to 3,300ft (1,000m) and stay submerged for an hour. Their internal organs are specially adapted to permit this, and they are also able to direct blood flow away from the body to the brain, and to store oxygen in the muscles. Their blood is very rich in haemoglobin and they have flexible chest walls to deal with the enormous pressures that such deep diving would subject them to. On surfacing the blow hole opens and the warm, moist, foul air is explosively expelled. The whale then takes in more air and dives.

Whales have a highly developed sense of hearing and the toothed whales in particular locate their prey – fish and squid – by sonar, emitting short pulses of ultrasonic sound and being guided by the echoes. It is not known how baleen whales locate their much smaller planktonic prey. It is also clear that whales communicate by a variety of noises; some of these are particularly low-frequency sounds which are audible through water scores of miles away.

Single calves are born after a gestation period of some ten to fifteen months and are pushed to the surface by the mother to take their first breath. They grow quickly on the mother's rich milk but may be dependent on her for a year or more and take several years to reach maturity. Most can only breed once in two or three years, and perhaps as seldom as once in six years for the sperm and killer whales.

Many species occur in British waters, especially in the summer months when they move north from the tropics to take advantage of the northern summer's flush of fish and plankton. These migrants return south in the autumn, some such as the blue, fin, minke and humpback whales covering some 8,000 miles (12,800km) on a round trip. They are most easily seen from headlands which command wide views of the sea, and are most frequent off the west and north coasts of Britain. The common porpoise and the bottle-nosed dolphin are probably the most frequently seen, and headlands such as Dungeness and Flamborough Head on the east coast are often included in their travels. Killer whales, too, visit the North Sea and the Farne Islands nearly always provide a good site in late summer. The largest whales appear to migrate along the edge of the continental shelf west of Ireland, and the best sites for most species are the headlands off the Irish south and west coast and, later in the summer, around the Outer Hebrides, Orkney and Shetland.

The immediate dangers to these superb, friendly, intelligent creatures are well known – men still kill whales, Norway, Japan and Iceland being the main nations involved. Nowadays, a new and more serious threat is the factory fishing of krill (a type of shrimp), an industry which is rapidly expanding as ordinary fish stocks become depleted in southern hemisphere waters. This is a situation that may threaten all the whales, seals and even penguin colonies of the more southerly latitudes, as it will result in a drastic reduction in their basic food supply.

The killer whale cruises towards inshore waters looking for prey which may be fish, seals or even dolphins. The killer whale is the largest of the dolphin family, and is easily recognised by its striking black-and-white body patterns and tall back fin. It is also among the most ferocious animals in the sea, but not generally aggressive towards man.

CHAPTER 9

MOUNTAINS AND MOORLAND

MOUNTAINS in Britain are usually composed of older, harder rocks; they have suffered millions of years of slow erosion so that now, many are just the worn-down stumps of what were once mighty peaks. Throughout the Ice Ages, the snow falling on the mountains accumulated to form glaciers, which merged to form ice-sheets, and it is these ice-sheets which sculpted the uplands into the shape we know and recognise today. Glaciers picked up great masses of rock and gravel and carried them along and these, together with the huge weight of the ice, gouged out enormous corries, or 'cwms', at the glacier heads. U-shaped valleys are typical of this glacial influence and were formed in all the mountain ranges; and water gathered in the scooped-out hollows, or where valleys were dammed by deposits of glacial debris, until it formed the corrie lakes and tarns that we see now. The valleys of the English Lake District are particularly notice-able when observed from the air, radiating outwards like the spokes of a wheel with the highest mountains at the hub.

Huge quantities of material were carried by the glaciers, sometimes many miles from their source, and were eventually deposited in vari-ous ways: as 'moraines' (lines of debris), or 'glacial drift' (flattened de-posits of boulder clay), or as vast plains of sand and gravel, all of which were left when the ice melted and the glaciers retreated. In the resul-tant meltwater rivers, great quantities of matter were washed into the valleys and lowlands, as 'fluvio-glacial deposits'. Such deposits contain many different types of rock, and certain differences in the shape and outline of these provide a guide as to their origin. For example, if the rock fragments are rounded and worn down, then they have been carried down by the rivers, and if the deposits consist of sharp, angular stones they have been left by a glacier.

Furthermore, the individual mountain ranges owe their differences of outline and character to the varying types of bedrock from which they are formed. Thus, the hard granites of the Cairngorms form a rounded massif quite unlike the serrated ridges of Snowdonia with its grits, slates and limestones. In Skye the Red Cuillins are made of granite and granophyre, and have been ground and weathered to smoother, more rounded hills than the rugged Black Cuillins which are formed largely of hard gabbro, with dykes of softer, more easily weathered basalt.

The process of erosion is continuous. Loose rock continues to break away from steep, unstable valley walls, adding to the scree slopes which over the centuries have accumulated beneath; streams wear the surface rock into channels which will ultimately become steep-sided gorges – where harder, less easily eroded rock is encountered, the water spills over as a waterfall. On the high summits the effects of frost and wind still shatter boulders into stony wastes.

There is evidence to show that some of those mountain areas above 2,300ft (700m) were higher than the ice of the last glaciation, and so managed to retain their relict arctic vegetation. Such places as Upper Teesdale come into this category, places which in prehistoric times were islands in a sea of ice.

Now the greater part of our mountains and moorlands are wide open and treeless. However, at one time forests covered much of upland Britain and it was man and his introduction of domestic grazing animals which together were responsible for today's wide open spaces. Initially the forests were cut down for timber and for fuel (the latter especially for iron smelting), and pastoral agriculture, based first on cattle and later on sheep, kept the uplands clear. During Georgian and Victorian times, management policy for many upland areas changed in favour of deer stalking and the shooting of red grouse, and this continues to the present day. More recently huge mono-cultured forests have appeared in many moorland areas in order to meet the demand for timber. Initially these were planted by the Forestry Commission for the state; now private companies participate, supported by substantial tax concessions and grants offered by the government.

Afforestation of the mountains and moorlands by dense plantations of alien conifers constitutes the greatest single threat to these last true wildernesses in Britain.

PREVIOUS PAGE
Slioch Mountain and Loch Maree on Scotland's west coast, a land shaped by Ice Age glaciers. The direction of Loch Maree, Loch Assynt and many lochs and streams in between indicate that the grain of the Lewisian gneiss on the western sea-board runs from north-west to south-east.

The Mountain Environment

Plants and animals have to overcome many problems if they are to live in the mountains, and in Britain the first is the cold, wet climate. It is fortunate that our shores are warmed by the waters of the Gulf Stream because the British Isles do in fact lie in the northern temperate zone – if it were not for the influence of the warm Atlantic currents our climate would be much more like that of Scandinavia, with long, cold winters.

The problems of wind, rain and cold are compounded by increasing altitude, as the tempera-

ture drops by 2°F (1°C) for every 500ft (150m) rise. The summit of Ben Nevis at 4,406ft (1,347m) has a mean July temperature of 43°F (6°C), while Fort William at its base, at an altitude of only 33ft (10m), has a mean July temperature of 60°F (15.5°C).

Rainfall increases considerably with altitude, and in Britain tends to fall on the west side of most ranges. A higher rainfall means more cloud, less sun, lower temperatures and less plant growth. Similarly, wind speed often increases with higher altitude, and there is far less shelter on the heights.

High winds have a significantly bad effect on plants, blowing away insulating snow cover in winter, driving millions of·sharp ice particles before them and generally stunting growth; a heather plant a mere 2in (5cm) high may in fact be quite

old in such conditions. Prostrate species have a clear advantage; on the highest summits low-growing plants like mountain azalea and least willow occur very frequently and máy become dominant. The least willow, in particular, is the supreme mountain shrub, with most of its twigs and branches underground. One old plant may cover several square yards but it will really only be noticed in summer when the young shoots appear, bearing flowers and leaves. Heather is found only up to an altitude of about 3,000ft (900m) in Britain, as above this snow tends to lie for more than six months of the year.

Geological difficulties add to those produced by poor climate. Most mountains consist of hard rock which weathers slowly and provides only a poor supply of nutrients. Soil tends to occur in pockets

Gordale Scar in the Yorkshire Dales National Park. Wet weather occurs very frequently on the high hills, rain tending to fall on the west side of most ranges.

221

between the bedrock, and water may not be able to drain effectively. Large areas may be covered in glacial drift, a thick impervious clay resulting from the long-term grinding of soil particles by the glaciers. Water drainage through these deposits is poor, giving rise to large areas of waterlogged ground.

Where the thick layer of glacial drift is absent and the underlying rock can provide nutrients, there is often a spectacular improvement in the vegetation, both in richness and in variety. This in turn has an impact on the number of insects, pairs of breeding birds, and mammals present.

Mountain vegetation and the animals that live on it are often clearly zoned. Perhaps the clearest line of demarcation is the tree line, near 2,400ft (700m) in the south, but sea-level in the extreme north. Trees provide shelter while above them lies a bleak, windy world. Above the tree line the vegetation may be described as montane, and the commonest type is stunted grassland, representing the average condition in the mountains, the sum of the effects of poor climate, poor soil and the uniformity produced from close-cropping by sheep and deer. High on the summits the frost breaks down the rocks into boulders and the first colonisers are lichens. The first plant to appear on these exposed tops is woolly-hair moss, more often called Rhacomitrium. Once enough soil accumulates from ground rock fragments and dead mosses and lichens, flowering plants may colonise, especially in sheltered spots between the stones. The rock wastes of the summits will appear full of the grey-green Rhacomitrium, and carpets of lichens with least willow, mountain sedge, and some heath bedstraw; small shoots of bilberry may also be present, and sometimes fir and alpine clubmosses, all with runners buried beneath the moss carpet and only the growing tips emerging.

Pollination in such a cold and wet habitat is clearly a problem. Insects need sunshine, and sunny days are rare. Most montane plants are thus perennials so do not need to produce seed every year, and many species have also developed vegetative methods of reproduction to supplement the sexual production of seeds. Some have even developed a viviparous system in which they produce small plantlets in the flower-heads. Plants such as alpine meadow-grass, viviparous fescue-grass and the very rare drooping saxifrage

all reproduce in this way.

The richest vegetation is found where rocks or scree occur, with running water to carry detritus, humus and weathered minerals into the soil. There are often small fertile areas such as this around the base of a rock outcrop, or along a stream bed; these are called 'flushes', and often contain the widest variety of mountain plants.

Montane grasslands are found on soils which are well drained, and are the most widespead vegetation type. The grass, usually sheep's fescue, is a natural succession from the rhacomitrium-heath type of colonising vegetation, and increases as the layer of humus accumulates. The principal factor in its development is now sheep grazing. Many areas have been over-grazed, besides which sheep tend to select the more nutritious grasses, but reject, for example, the mat grass Nardus which then grows indiscriminately. Large areas of the uplands have deteriorated because of over-grazing.

Where the soil is over 9in (23cm) or so in depth, the lower slopes are likely to have areas of bracken. Bracken has spread enormously in the last 100 years and as it shades out the more edible grasses it renders the land almost useless as pasture. It will only grow to an altitude of about 1,000ft (300m); on the leached soils above this level mat grass is most likely to be dominant, with bilberry and heather. The true heather moor is largely man made, and the management policy of successive burning over many years has helped it spread and thrive.

For an animal, life in the mountains is harsh and few species apart from invertebrates have adapted to the rigorous demands of the mountain environment. Survival through the winter is the main test and most spend it in an inactive state; insects, for example, overwinter as eggs or pupae buried in the soil. Hardly any large animals or birds can cope with the severe climate, and most will migrate for the winter months. Some, like the red deer, may simply move to the sheltered lowlands; others like the dunlin and the golden plover may travel hundreds of miles to food-filled estuaries. The ptarmigan and the mountain hare are the best mountain survivors, and have adapted to cope with winter snows, but in Britain's cold northern mountains even these hardy specialists may have to seek refuge on lower ground in bad weather.

Moorland, Bogs and Peat

Open dry moorland, blanket bogs and peat occur typically over acid soils, and are very similar in structure and characteristics. Thus dry moorland results where an old blanket bog has dried, or on better-drained soils – and if the dry moorland became waterlogged it would slowly change to blanket bog; drainage is therefore the crucial factor, and whether these areas become bog or dry moorland relies entirely on the extent to which their constituent structure is drained. Both habitats occur between the lower enclosed farmland and the higher mountain crags, from about 1,000 to 2,000ft (300 to 600m). This height represents the

upper limit of moorland, and above it true montane conditions prevail.

Moorlands are strikingly open, windswept and desolate places; houses and roads are few and far between, and there are hardly any trees. At one time, however, most of the moorland was covered in forest. Evidence of this can be seen in the sides of gulleys or where peat cutting is taking place for fuel, as the well-preserved remains of ancient trees, and certainly the seeds from them, are often revealed buried in the peat. It is therefore entirely due to man's influence and management policies over the years that the moors are now so open, and by grazing the hills with sheep and burning off the older, taller moorland growth, he keeps the hills open.

Most of these moorlands have at least 60in (1,500mm) of rain per year. Where the slope and drainage is good, the bulk of this water will run off

Polytrichum, a striking moss which grows in large clumps among bogs. Afforestation is the greatest threat to upland bogs as, if the peat is ploughed, drained and fertilised, this prevents the accumulation of any more peat and the natural cycle is broken.

over the rock as streams and burns, and the boulders will often be draped in mosses, liverworts and lichens. Further evidence of the former woodland cover lies in the fact that essentially woodland plants such as primroses, wood anemones and ferns shelter in the steep rocky banks.

There are, then, three different types of soil/peat covering which support three main types of moorland vegetation: the blanket peat bogs which form in cool, permanently wet areas, and the two dry-ground habitats where grass moorland and dwarf shrub heath predominate. These last two are maintained primarily by the hand of man, the grass because it is constantly grazed by sheep, and the heather because it is burned off to precipitate regeneration, the new shoots being the basic food supply for grouse. The acidity of the rock means that only a few species of plant manage to thrive; dwarf and shrub heath is dominated by common heather or ling, and to a lesser extent bilberry. The latter, along with the deep purple bell heather, is particularly conspicuous as an early coloniser of recently burnt heather moorland; bell heather is usually found on the most sandy, dry parts of the moor. Where there are boggy pools the pink cross-leaved heath is the dominant plant of the heather family.

Before the forests were cleared, heather comprised no more than a small part of the forest-floor vegetation, mainly in clearings where the sun could reach it. However, once the trees were removed heather became dominant, especially in the drier eastern Highlands of Scotland, Yorkshire and the eastern Pennines. These are now the prime grouse moors, and a great deal of research has been applied to the biology and ecology of the red grouse. The single most important factor for grouse is a good diet of young heather, and this is produced by a systematic burning programme. Left to its own devices, the grouse population tends to fluctuate on a natural cycle, but it can be doubled by skilful management of the burning programme. Heather may live for as long as forty years, but after about twelve to fifteen years it becomes old and woody. If at this stage it is carefully burned off, regeneration will provide a quantity of new shoots; and if only sections of the moor are burned at a time, it is possible to provide a constant supply of rich new shoots for the grouse. It is

also important to leave strips of old, dense heather for nesting cover and shelter. Not many of the tourists passing through the Highlands ever realise that the glorious patchwork of purple heather is entirely man-made.

The grass moorlands have the poorest flora and fauna, primarily because of the huge numbers of grazing sheep. Sheep will nibble nearly anything except bracken and mat grass, and they effectively prevent any regeneration of woodland. Birch trees, for example, live for some eighty years, but the seedlings in a birch wood regularly sheep-grazed will not survive, so the wood will thus vanish completely in eighty years without a tree being felled. Flowering plants only survive in any numbers in places inaccessible to sheep, between rocks and boulders and on cliff ledges. The flora of grass moorlands therefore includes widely scattered yellow tormentil, white heath bedstraw, blue milkwort and pale blue common speedwell. The birds most commonly found are the meadow pipit and the skylark.

Bogs provide a much richer habitat than the dry moorland, although the flora and fauna is still very restricted compared with lowland fens, for example; a rich lowland fen is on alkaline, base-rich soils, whereas an upland bog is an acid, and therefore less productive environment. Bogs form on poorly drained upland soils, often where there is a thick underlying deposit of glacial-drift clays left by the last Ice Age. They are called blanket bogs because they can literally blanket an area and its vegetation, including its trees – trees subjected to constantly acid, waterlogged conditions will eventually die and be swamped by the bog.

One group of plants however, thrives in these conditions: the bog-mosses, or sphagnums. There is little decomposition in acid water and a new growth of bright green bog-moss inevitably occurs on top of the old dead material. As the process con-

Cotton grass is really a sedge and is one of the most characteristic plants of blanket peat bogs. It flowers in May and June, and the distinctive long cottony hairs are attached to the fruits which develop later in July.

tinues so the layers are compressed and peat is formed, in some places to a depth of 15ft (5m). Bogs only survive if they are wet all year round. Once they dry out, or if some form of drainage is introduced, the sphagnum is replaced by 'dry' plants and the peat itself may erode.

Bog flowers are often very specialised, though restricted in variety. Several plants compensate for the lack of nutrients in the soil by being insectivorous and trapping their own supply; for example, the sundews and butterworts have rosettes of specially adapted leaves covered in sticky filaments which will trap an unwary insect and close over it digesting the chemicals from its body. Bladderwort is another insect-eater – found in bog pools, each stem possesses a flask-like bladder with trigger hairs at the entrance; these will be 'sprung' by tiny aquatic animals which the plant then sucks into the bladder. Perhaps the most characteristic plant of blanket peat is cotton-grass, or cotton-sedge; in summer its white, cottony flowers cover huge areas in a waving carpet. It supports many insects which themselves provide food for the numerous meadow pipits and for the waders: dunlin, curlew, golden plover, redshank, snipe and greenshank, all of which use upland bogs as prime breeding sites.

Birds of Crag and Mountain Top

Only three species of bird in the British Isles can be regarded as truly montane: the ptarmigan, dotterel and snow bunting. All three breed in the short northern summer on the stony summits of the highest British mountains, usually in Scotland, although in the last few years the dotterel has certainly bred in Wales and Cumbria.

The ptarmigan is a high-mountain grouse and is resident all year, although it may move to lower moorland in winter where it then competes more directly with its close relative the red grouse. It moults to a white plumage in winter which provides superb camouflage in snow, and its legs, too, are covered in dense white feathers which helps to reduce heat loss. Ptarmigan feed on the leaves, shoots and berries of bilberry, crowberry and heather, and will also eat insects when these can be found. They are hard to spot in their world of broken rocks and lichens, and are not usually noticed until they are actually flushed, when they whirr away like bullets on rapidly beating white wings. The population is estimated at 5,000 pairs. The female produces one brood a year of five to nine eggs; incubation is twenty-two to twenty-six days, and the nidifugous chicks fledge very early, at only ten days, before they are half grown. Chick mortality is high in bad weather.

The dotterel nests on the highest tops in Britain, and now there are over 850 pairs. A small, migratory wader of the plover family, it arrives in early May from Africa, and it is the male, less brightly coloured than the female, which incubates the eggs and tends the young.

The delicate pied snow bunting is an Arctic bird; it is on the southern edge of its breeding range in Britain with fewer than twenty pairs nesting among the lichen-covered boulders of the highest summits.

The king of British birds must be the magnificent golden eagle, and it is the eagle which is more widely thought of as the classic mountain bird. Usually, however, it breeds and often hunts at much lower altitudes around the tree-line at the 2,000ft (650m) level. Pairs mate for life, and ring the changes on three or four eyries, on steep crags or in tall Scots pine trees. Nests are often placed on a stunted tree growing from a rock cleft in a high cliff face, and are added to over the years, becoming huge platforms of twigs and branches which may be 10ft (3m) high and 6ft (2m) across. Nest restoration may start in March, when the mountains are still snow covered; two eggs are laid in

Ptarmigan nest high in the mountains; the scrape, scantily lined with grass or heather, is usually in an open situation but sheltered by a stone or mound. The eggs are laid in late May. Both sexes have a red wattle over the eye, larger in the male.

early April and hatch in six weeks, three to four days apart. Where food supply is good both chicks may survive, but the younger and weaker chick often dies, and this is now regarded as an extreme form of natural selection in a harsh mountain world where food is usually sparse. The eaglets fly after some nine or ten weeks and stay around the nest site for several weeks afterwards, flying with their parents and gaining experience of their environment.

Eagles are often confused with the much more numerous buzzard, but the eagle has a longer, more prominent head, and a slower, more powerful wing-beat. Its wing-span reaches some 6ft (2m) and it is about 3ft (1m) long, the female being a couple of inches (5cm or so) larger than the male. There are approximately 450 pairs of eagles to 12,000 pairs of buzzards in Britain.

All sorts of food appears on their diet, from voles to greylag geese. They eat carrion, especially in winter, feeding from the carcasses of deer and sheep which have died in the snow. A few living lambs are undoubtedly taken but research has shown that this is not nearly as frequent as some hill-farmers believe; most lamb carcasses found at eagle eyries have been picked up dead on the ground.

For centuries man has used the eagle as a symbol of power and esteem, but in spite of this it has been persistently persecuted, especially by ill-informed gamekeepers and hill-farmers. Eagles are something of a tourist asset in the Scottish Highlands, and thousands of visitors travel there in the hope of seeing their first golden eagle.

The peregrine, too, is a bird of the crags; in the 1950s and 60s its population declined to a critical level because of pesticide pollution, and it disappeared from many of its customary haunts. The population fell to about 200 pairs as organochloride pesticides (DDT, lindane and dieldrin) worked through the food chain from plant to insect to bird. Not only did eggs fail to hatch but their shells were abnormally thin, and many were smashed during incubation. These chemicals were withdrawn from agricultural use in the late 1960s and, twenty years later, the peregrine has recovered to about 900 pairs and reappeared in many of the places from which it had vanished.

The peregrine has one brood a year, usually of three to four chicks; incubation lasts about four weeks and fledging a further five to six weeks. The adult bird is a masterpiece of evolution, supremely powerful and has a mastery of the air which few other birds possess, even though it may be slower in level flight than some of the birds it preys on. Most pairs nest on high rock faces, but they are nonetheless vulnerable to man's acquisitive nature – they are still regarded as the prince among falcons by falconers the world over, and fetch high prices when smuggled illegally to the countries of the Middle East.

Also occupying the high mountain fastnesses is the raven, more frequently encountered in high hills throughout northern and western Britain. It breeds on rocky crags from late February, and is often the earliest of Britain's birds to be sitting on eggs, perhaps even amidst the ice and snow of late winter. Ravens, like eagles, need large areas of open country in which to seek their food, and there is evidence that the widespread afforestation of the uplands has caused a substantial reduction in their numbers since 1972 when the population was estimated at 5,000 pairs. There is one brood, of three to four chicks; incubation takes three weeks and fledging five to six weeks.

Ravens are the 'vultures' of sheep country, and patrol the uplands in search of dead or dying sheep – although they will eat anything from acorns to washed-up shellfish. They are easily distinguished from other crows on the hills by their large size, 'bull' head, wedge-shaped tail and deep, croaking call.

Generally, the restricted amount of food available in the mountains limits the population densities of all bird species, but in the short summer months insects may be abundant and this enables meadow pipits in particular to live up to the 4,000ft (1,300m) level. Meadow pipits are usually the most numerous passerine species above 1,500ft (480m), with an average density of 40 pairs per square mile (20 pairs per square kilometre). This may be only 20 (10) pairs on poor-quality moorland but as high as 80 (40) pairs in new, drained forestry plantations where the ground cover is dense. The meadow pipit constitutes a major source of food for moorland birds such as the merlin and hen harrier, and therefore plays a vital role in the ecology of the mountains. Next in importance are skylarks at about half the density of the meadow pipits, although they are much more numerous on grassy moors.

As climate and migration patterns change slightly, so more Scandinavian and Arctic species may occur. Here and there on the highest places there may even be a pair of shorelarks, Lapland buntings or purple sandpipers, all species of the far north and all recent colonists.

Birds of Moorland

Between the valley woods and the high, cold mountain-tops lie the open moors. These extend throughout highland Britain, an extensive, wet, windswept landscape dominated by peat-covered plateaux, and by tarns, pools and bogs. In winter this habitat is very bleak and few birds can tolerate its shelterless expanse. In spring, however, there is a sudden rush of returning birds back to reap the moors' harvest of abundant food, unlocked by the warmer season for just a few short weeks from April through to July. By early autumn most of the birds will have vanished, leaving the moor quiet except for gunfire when the annual grouse shoot be-

The peregrine population has recovered considerably since the level fell drastically due to the use of pesticides in the 1950s. The male, the tiercel, is smaller than the female and has less heavily barred underparts. The power dive of the peregrine as it swoops on its prey is a stupendous sight.

gins. Spring shows moorland at its most attractive; curlew and golden plover call across the hills, snipe drum over the bogs and grouse chatter in the heather.

Vegetation falls into three categories: heather shrub, blanket bog and grass moorland, all usually on acid soils of low fertility. Where this fertility increases the population of breeding birds is much higher, as can be seen in some parts of the Pennines. Open grassland is the least productive of these three habitats and only the ubiquitous meadow pipit and skylark may manage to breed successfully. Where a small outcrop of rock ap-

pears in the grass, so too may a pair of wheatears, nesting among the holes and crevices. The blanket bog and heather shrub are occupied by three groups of birds: waders, game birds and predators, although the meadow pipit and wheatear may be found wherever suitable cover exists. A separate habitat, that of open water, attracts several other notable breeding birds.

Heather moor is dominated by common heather, or ling, with bilberry and purple bell heather intermixed on drier ground; it occupies vast areas of the uplands. Much of this is managed by man, sections of it burnt off at intervals to provide new heather shoots for red grouse, the prime game bird of the moors. Heather moorland would vanish in many areas if it were not for the sporting interests of the grouse shoot. The red grouse is sufficiently hardy to spend the whole year on the moor, occupying the zone between the black grouse of the lower moorland edge and the ptarmigan of the montane zone above 2,500ft (750m). It is now regarded as a sub-species of the more Arctic willow grouse of the northern forests.

The red grouse feeds mainly on heather shoots with cotton-grass flowers, berries, seeds and insects, and its population fluctuates in a more-or-less regular cycle over a number of years, in common with many northern species. Artificially high numbers can be induced by controlled heather management, when as many as 100 pairs per square mile (50 pairs per square kilometre) may be found. It is a strictly territorial bird and the final population is governed by the number of territory-holding males in residence. The total population is between 200,000 and 400,000 pairs. The female produces one brood of six to eleven chicks which can fly at about twelve days old.

Where the ground is wet, peat-bog vegetation provides a major breeding habitat for waders; its bog-mosses, cotton-grass, heather, cross-leaved heath and deer sedge, interspersed with wet hollows and pools, supply ideal nesting sites. Nine species breed regularly, although the red-necked phalarope is rare, confined to tiny lochans (small lakes) in moorland on the Scottish outer islands. There are probably only about twenty pairs in Britain. The whimbrel, too, is almost entirely confined to the northern isles, although like the phalarope, one or two pairs may breed in the remote bogs of Caithness and Sutherland. The total British population is about 460 pairs. Wood sandpipers and Temmink's stints are even rarer colonists from Scandinavia which might occur in the far north.

The lapwing is the most familiar of the upland

bog waders, although it is much more likely to frequent farmland at lower altitudes, where it breeds on rough pasture and arable ploughed fields. Like most waders it lays four eggs, and the chicks, which hatch after a month's incubation, are able to leave the nest within hours. Like all ground-nesting species, the lapwing has superbly camouflaged eggs and chicks to reduce predation in the open habitat.

Curlew return to their breeding moors in early spring having spent the winter on estuary mudflats, the males filling the moorland air with their attractive territorial displays and song. The bird sitting on its clutch of eggs is beautifully camouflaged, though the off-duty bird will still keep look-out, calling a warning at any sign of danger. There is one brood a year consisting of four nidifugous chicks, able to fly at five to six weeks old.

The golden plover, like the curlew, returns to the moors in spring. It is a more upland bird though, breeding on the higher, flat plateaux from 1,500 to 3,000ft (500 to 900m).

Where the ground is particularly wet there is always a chance of finding a group of the tiny, smart, black-chested dunlin, the least numerous of the regular upland waders.

National wader populations in 1986 were estimated at 9,150 pairs of dunlin, 181,500 lapwing, 35,000 curlew and 27,400 pairs of golden plover, although most of the lapwing were on lowland pastures. Like the golden plover, few dunlin breed south of the northern Pennines, although a tiny colony still exists on Dartmoor, with fourteen pairs in 1986, the most southerly colony in the world. Britain is at the southern limit of the dunlins' predominantly Arctic breeding range.

Snipe and redshank, more often associated with lowland, fresh-marsh fields, are fairly frequent breeding birds of the wetter moorland areas, snipe especially occurring in scattered colonies. These small birds, with their evocative drumming display flight, present a dusk-time spectacle rivalling the display of curlew and golden plover.

Finally there is the rare greenshank; it has a population of about 1,200 pairs and is a notable inhabitant of the northernmost moorlands in Scotland, arriving from Africa by late March.

Where bog drains into moorland tarns and pools, rare wetland birds are found, mainly in the Highlands of Scotland. Black- and red-throated divers are typical of the habitat, the red-throat more often on the tiny hill lochans; the black-throated diver, now very rare, is found more frequently on the larger lochs of the bleak north-western valleys. Slavonian and black-necked grebes are found on hill lochans in a small area of the central Highlands. Where the moor is dissected by stone walls or runs into broken scree, perhaps with heather scrub, stonechat, whinchat, wheatear and twite may all breed, although the twite is a species which is declining, and is now restricted to the south Pennines and the Scottish moors.

The predators reign supreme over all these areas. Eagles and peregrines hunt down across the moors from their craggy eyries, while the hen harrier, the merlin and the short-eared owl are the classic moorland predators; they all breed in the heather. However, the merlin population is in serious decline, probably now only half the 1972 estimate of 600 pairs. Hen harriers total 5–600 pairs and the short-eared owl 1,000 pairs in poor years and maybe 8–9,000 pairs in peak years. The latter's breeding is geared so tightly to the population of its main food, field voles, that it can produce many more eggs in the best years. Harriers take birds as well as voles, stalking them with a slow, low, gliding flight. Merlins, on the other hand, dash at high speed across the heather to pluck a pipit from its perch. The decline in merlins is probably due to a loss of habitat from afforestation and over-grazing. Hen harriers and short-eared owls have increased in the past twenty-five years, taking advantage of the low, thicket stages of new plantations; once the trees gain height, however, their populations will decline. Like the waders, most of the birds of prey retreat in winter to the estuaries where there is more food.

The merlin is Britain's smallest falcon, barely larger than a blackbird. Its population has fallen from 1,000 to 500 pairs in ten years. It flies fast and low on narrow pointed wings, hugging the contours of the land. All the merlin's agility and manoeuvrability are apparent when it chases the small birds which form the major part of its diet.

Birds of Upland Stream and Forest

Below the crags and scree slopes of the high summits lies a harsh world of windswept, rain-soaked moorland, ancient moss-covered woods and 'new' man-made forests, criss-crossed by innumerable rocky gulleys and streams. Here and there on level ground or in a hillside basin a mountain tarn or loch may form, or peat bog accumulate. Along the streams and in the woodlands are many birds specific to these upland habitats.

The most adaptable of the small riverside passerines is the grey wagtail, dependent upon running water for its insect food. The dipper is also restricted to running water, but whereas the wagtail may occur on very small streams at high altitude, the dipper is found on larger streams and boulder-strewn rivers. While the grey wagtail catches flying insects, the dipper feeds underwater on the bed of the stream.

Another typical streamside bird of the uplands is the common sandpiper, found in many areas above 1,000ft (300m). It feeds by carefully picking items off the stream edge, and thus does not compete with the other species. It is strongly migratory, leaving for warmer climes by midsummer. The

total British population of the common sandpiper is estimated at 20,000 pairs, most of these being found along upland water-courses. It arrives from Africa and southern Europe in late spring.

There are two other waders which in spring are very likely to nest along the banks of the upland rivers, or more frequently among the stones on winter storm beaches: the oystercatcher and ringed plover. The ring ousel is quite numerous along the gulleys and stream-cut ravines in upland areas from Dartmoor to Cape Wrath. It arrives in Britain in March and April, having spent the winter in southern Europe and North Africa. It is closely related to the blackbird, and produces one or two broods in a blackbird-style nest on a rock ledge or in thick heather, before returning south in August and September. The population is estimated at between 8,000 and 16,000 pairs.

Where water and forest meet, several species of northern ducks may occur. The goosander and red-breasted merganser are the most widespread, the merganser favouring the upper reaches of the rivers, while the goosander is more frequent on the lower, slower stretches in the wooded foothills. Britain's population of mergansers probably numbers 2,000 pairs and is found mainly in Scotland, but some now breed in the Lake District and in Wales too; the 1,000 pairs of goosander are also found in Scotland. Both species eat small fish and breed in thick vegetation not far from water, mergansers on the ground but the goosander using tree holes. The lovely piebald goldeneye, a rare colonist from Scandinavia, also uses tree holes, and its numbers are increasing along Scotland's rivers and lochs largely due to the provision of judiciously placed nest boxes by the Forestry Commission and the Royal Society for the Protection of Birds. Regular breeding of the goldeneye started in Britain in 1970 with only one pair, but

A member of the duck family, the goldeneye is primarily a winter visitor from Scandinavia, though approximately 100 pairs now breed in Scotland. The goldeneye is extremely active, diving constantly in search of its diet of molluscs and crustaceans. During the mating season females are indulged by the male display of throwing the head back and raising the breast. In both sexes, the peaked crown and short bill give the impression of a triangular-shaped head.

the population is now up to about 100 pairs. All three species move to coastal areas in winter.

Mallard, teal and tufted duck breed widely in the uplands, and most of the wigeon population – a few hundred pairs – is found in Scotland. Natural upland woods are usually of Scots pine or sessile oak, the pine being largely Scottish and the oak woods in the Lake District and Wales. However, man-made forests of alien conifers now cover a much greater area than natural forests, and birds such as crossbills and goldcrests which live in the tree canopy have successfully made the transition to these new plantations; scrub- and ground-nesting birds have not.

Several birds of prey and owls live in Britain's upland forests. The sparrowhawk and the tawny owl are the most widespread, with the long-eared owl more sparsely distributed. The buzzard, too, is a surprisingly common bird in the hanging valley oak woods of the West Country and Wales; it sometimes nests in crags, though will more often use high forest trees. Population estimates show a fairly stable figure of some 12,000 pairs.

They all occupy their own distinct ecological niche. The buzzard feeds away from the forest on upland hares, rabbits and voles; the sparrowhawk hunts small birds swiftly in flight through the daytime woods. The tawny owl hunts rodents at night within the boundaries of the deciduous wood, while long-eared owls prefer conifer woods and hunt by night over open moorland. The tawny owl uses tree holes while the long-eared takes over old hawks', crows' or even squirrels' nests, high in the conifer branches. The estimated tawny owl population is 50,000 pairs, and that of the much more elusive long-eared owl, 3,000 pairs.

The goshawk and the fish-eating osprey are two large and rare birds of prey whose normal habitat is mountain forest, but only some fifty pairs of each have so far colonised Britain. The number of red kites, too, is a tiny remnant of what was once a much wider population; there are estimated to be just fifty pairs now, found in central Wales.

The upland woods come alive in the spring as birds flood back to breed, some from as far away as Africa; others, like the crossbill and siskin, may not have travelled further than the lower riverside woods. Where the woodlands are old and full of glades, clearings, rotting stumps and a dense scrub

layer, the smaller passerines are often numerous despite the colder mountain climate. Chaffinches are often the commonest breeding bird, whilst redpolls have increased in birch woods. Wood, willow and grasshopper warblers, redstart, spotted flycatcher and blue and great tits are found in ancient pine and oak woods; the pied flycatcher is largely a bird of the valley oak woods of the west coast. Ancient pine woods in Scotland have the rare crested tit, along with coal tits, goldcrests, siskins and crossbills – all pine specialists. The populations of the crossbill and siskin fluctuate with the abundance of the pine-cone crop.

Down on the forest floor both capercaillie and blackcock occur, although the latter is found more often on the moorland edge. Both these species are more or less confined to Scotland. Blackcock are found in the Pennines but have declined almost to vanishing point, while the Exmoor blackcock have become extinct.

A much more frequent ground-nesting bird of upland woods is the woodcock, a wonderfully camouflaged wading bird perfectly adapted for life in these damp forests. It uses its long bill to probe for food in the wet and boggy areas, and nests among the leaf litter. In the evenings in spring the woodcock will come out and fly the rounds of their woods at dusk, a practice known as 'roding'.

The tawny owl is widespread throughout Britain in all types of woodland. A nocturnal bird, it roosts by day in a hollow tree or close against the tree trunk. Its prey, mainly of wood mice and bank voles, is mostly taken from the ground though some birds such as starlings and sparrows are also taken from their roosts.

Mountain Flowers and Insects

The mountains of Britain are among the most exciting places in which to search for flowers, and the remoteness of some of the high Scottish peaks and glens will invariably make the modern enthusiast feel as much a pioneer as the early botanists. There are still rare alpine plants to be discovered in these high, wild places.

The most widespread of the mountain habitats is montane grassland, known as festuca-agrostis grassland because it consists largely of sheep's fescue and fine bent grasses. The flora is poor, its plants widely scattered and found in nearly all areas; these include: heath bedstraw, tormentil, devil's-bit scabious, common milkwort, common speedwell, lousewort and sheep's sorrel.

Heath vegetation covers large areas of the drier parts: heather is dominant, along with bilberry and bell heather. These heather moors are largely managed for grouse shooting.

The most exciting mountain flowers are usually found in 'flush' vegetation, along streamsides or below rock outcrops where extra nutrients have been washed into the soil. Most British mountains are made of hard, nutrient-poor rock, but where bands of limestone or lime-bearing rocks occur at or near the surface, the mountain vegetation of 'arctic-alpine' plants may improve dramatically, sometimes whole mountainsides being covered in alpine plants.

Ben Lawers is perhaps the most famous botanical site; great stretches of its western slopes are covered in dense growths of alpine lady's mantle. Ben Lawers and the other mountains in this range are composed of mica-schists, full of lime, which have transformed the upper terraces of these hills into hanging alpine rock gardens. A well run and informative centre belonging to the Scottish National Trust has been set up on the southern lower slopes of the mountain. Sheep grazing and moorland burning have had a significant effect in this area, and the mountains are largely treeless; deer sedge, purple moor-grass and heather abound in the lower areas.

High rainfall means that the lower, flatter areas are very boggy, and the classic flowers of peat bogs will all be found: bog myrtle – a low shrub – abundant cross-leaved heath and sheets of bog-mosses or sphagnum. Rooted in the bog-moss carpets are round-leaved sundew, butterwort, bog asphodel and marsh violets.

Where the ground is slightly drier, there will be the common grassland flowers dotted among the heather and bilberry: tormentil, lousewort, milkwort and heath bedstraw. Dips and hollows which collect run-off water have growths of common rush, sharp-flowered rush, star and common sedges and cotton-grass. All these plants can be seen on lowland English heaths.

Once up at the 1,500ft (480m) level, any rocky outcrop on Ben Lawers will support alpine plants. The tiny alpine meadow rue is likely to be abundant on wet flushes, with viviparous bistort; yellow saxifrage occurs widely in these flushes, too. In early spring, purple mountain saxifrage adorns many crags and outcrops. This hardy arctic plant grows as far north as 84° latitude, and forms its buds beneath the winter snows. The little streams or 'rhylls' are filled with plants, especially yellow mountain saxifrage and alpine lady's mantle. Common wild thyme often accompanies them among the flaky lime-filled rocks.

The cliffs and screes on Ben Lawers will have a score and more of alpine plants including mossy, yellow and purple saxifrages, moss campion, alpine scurvy-grass, mossy cyphel, northern yellow rattle and alpine mouse-ear; quite apart from the more common lowland plants such as frog orchid, butterfly orchids, moonwort, grass of Parnassus and kidney-vetch. Globe flower, roseroot, holly fern and northern bedstraw are all found here, and on any of these mica-schist hills of the Breadalbane range. Wherever base-rich rocks such as limestone or basalt occur in mountain districts so a varied crop of montane plants may flourish.

Some plants on Ben Lawers are extremely rare –

Moss campion is an arctic-alpine plant found on the highest mountain ranges in Britain. Its beautiful deep rose-pink flowers can be seen from June to August.

Although normally widespread on cliffs and rock-ledges in mountains, the moss campion is also found at sea-level in the north of Scotland.

plants such as drooping saxifrage, snow gentian, alpine fleabane and mountain sandwort are found hardly anywhere else. Several dwarf willows grow here too: whortle and downy willows, mountain and woolly willows, all of these are rare, ground-level shrubs.

Further north, the Clova mountains east of the Braemar–Glenshee road have lime-rich rocks similar to those of Ben Lawers and are equally rich in alpine flowers, including species such as yellow oxytropis and alpine sowthistle not found at Ben Lawers.

The high Cairngorms are largely of granite and consequently have a less rich flora, but two species which are special to this area are arctic mouse-ear and tufted saxifrage.

Many arctic flowers occur at a much lower altitude. For example, on the Moine series of rocks from the Fannich Hills to Beinn Dearg the soil is calcareous and the flora correspondingly rich, with several saxifrages, moss campion and many more.

Inchnadamph and Knockan have massive outcrops of limestone, and these are responsible for a rich green landscape in an otherwise austere country of quartzite and gneiss. Mountain avens and whortle willow are found in great abundance, along with dark red helleborine. In this far northern corner of the Scottish mainland, alpine plants grow at or near sea-level – many species occur here, fully 1,000ft (300m) lower than in the central Highlands. This is most apparent from Durness round to Bettyhill where mountain avens, purple saxifrage, moss campion and purple oxytropis grow in remarkable profusion on shell-sand terraces on the coast.

In the summer months, insects are abundant in the uplands, but the number of species present is very much lower than in the lowlands because of the extent to which the climate and largely acid soils restrict plant variety. Three butterfly species are typical of the hill-country, all of them browns; they feed on grasses and sedges and produce only one brood of larvae because of the short upland season. All overwinter as larvae, buried deep in the grass tussocks. Thus: the small mountain ringlet is a truly montane butterfly found on grassy mountaintops in late June and July, above 1,800ft (575m) in the Lake District, and from about 1,500ft (480m) in the central Highlands. The large

heath is widely distributed in the north and west, as far south as Cardigan, flying for four to five weeks from mid-June; while the Scotch argus is a more northern butterfly but quite common on hills from Yorkshire northwards from the end of July and throughout August.

The moths, flies and beetles of upland regions can also be very numerous but again, they are always poor in variety. Many moths feed on heathers, and some of them are spectacular, such as the large Emperor moth, the northern eggar, the fox moth and the drinker moth. Some moths which are nocturnal in the lowlands have adapted to daytime flight in the mountains, taking advantage of the higher temperatures.

The vast summer glut of midges and craneflies attract many lowland birds on to the mountains to take advantage of the food supply; rooks, jackdaws, swallows and martins may suddenly appear in hundreds over high mountain slopes. They join the insectivorous meadow pipit, which, because of the abundant insect supply in summer, is the most widespread mountain bird.

Mountain avens is found near to sea-level in the far north of Scotland, and also grows on high mountain slopes farther south. Its dark-green leaves are white-felted underneath and the white flowers are seen from May to July.

Mammals of the Uplands

Several species of mammal are found almost exclusively in hill or mountain country; and other species which are better known in the lowlands also extend well into the mountains, like the fox. The herbivores are the easiest to see and most of them are about during the daytime, unlike the predators which tend to be more nocturnal.

The mountain hare is an animal solely of hill country, except in Ireland where there are no brown hares and hence no competition; thus in Ireland it is found right down to sea-level, and keeps its brown coat all year. In Scotland the mountain hare is native to the Highlands, but it has been introduced to the southern uplands, the Lake District and the Peak District. It usually moults into a white coat in winter which provides excellent camouflage especially in patchy snow and boulder fields; even the fur on the soles of its feet turns white, leaving just the black ear-tips which are almost impossible to distinguish in a snow-covered landscape. The occurrence of the white coat is rather variable, and depends upon altitude, temperature and age.

Mountain hares are easiest to see in the central Highlands, especially on Speyside and Deeside. In this area their population may reach fifty to seventy per 1,000 acres (400ha) – compared with no more than ten in the Lowlands and only one or two north of the Great Glen – although these figures may be much higher in years of peak population. They feed on grass and heather, the latter making up 90 per cent of their winter diet in snow conditions.

Red deer spend some eight months on the hills, only retreating to lower ground in the face of winter snows. Here they come into direct conflict with foresters and farmers because they eat their crops. They were originally driven up into the hills when their natural habitat, the lowland forests, were felled, and also because of hunting in days gone by. There are estimated to be some 270,000 red deer in Scotland. Natural regeneration of the trees is prevented largely because of the millions of sheep, but the deer are sufficiently numerous to

be of influence too, and their numbers necessitate the erection of deer fences in commercial forestry operations.

In summer, the Highland red deer live high on the hills, seeking out the coolest places – on long-lying snowdrifts or cool gulleys – to avoid flies. They may also descend to lush streamside pastures if disturbance is minimal. The stags usually pass the summer in small groups of ten to a dozen, but the hinds (females) gather in early June in traditional calving areas in sheltered mountain valleys, and each gives birth to a single calf. These large herds of hinds, calves and yearlings break up in late summer as the stags become more aggressive towards one another, and the rut starts in earnest in late September or October when each stag tries to gather a group of hinds for mating.

The red deer is the largest British land mammal; the average Scottish stag weighs about 15 stone (95kg), while a stag from a lowland wood, where better food is available, may be 30 stone (190kg). A pair of antlers may be 3ft (1m) long and weigh 15lb (7kg). On average a deer lives for five years, but they can live for twenty.

Roe deer are frequently found in many hill areas and are common in the Highlands south of the Great Glen. However, they are really forest animals preferring to live in the woods all year round, drifting out to feed in the fields or woodland edges early and late in the day. They live singly, or in a single family group; the rut is much earlier in the summer than with the red deer, and no herding takes place. Although mating is in July and August, delayed implantation of the foetus ensures that the fawn is not born before the following May.

Small herds of wild goats are found in many hill districts, with some seventy separate groups in Scotland alone. These are 'feral' – descended from past domestic stock which escaped and became wild. They cope well with the harsh conditions in the hills, and have become smaller, and carry a thicker fur than modern domestic goats.

The dominant small mammal of the mountains is the short-tailed or field vole, although tiny shrews have been found on Ben Nevis summit, and wood mice live on scraps on Snowdon. The thick, tussocky grassland provides the vole with ideal conditions, and these are improved even more in the early stages of new forestry when the ground is drained. In these new plantations field voles may reach very high numbers, providing a bumper food supply for birds of prey such as hen harriers and short-eared owls, which may rear larger than average families as a result. However, if vole numbers reach plague proportions, they can cause immense damage to new plantations by stripping bark. They are very hardy, and are frequent even in the rhacomitrium-heath zone of the hill summits.

The pygmy shrew is also found on moorland, in greater numbers than the common shrew because of the type of food available – many small spiders and beetles but few earthworms.

Field voles must be the staple diet of most of the predatory mammals of the hills. The fox is chief among these, and where food is sparse its hillside home range may be as much as 2,500 acres (1,000ha) in extent – compared with only 100 acres (40ha) in a city suburb. Foxes are found, thinly scattered through the hills, from Cornwall to Caithness and are most frequent in forestry plantations with a high vole population.

In the Highlands the fox must compete with the Scottish wildcat – very wary and very shy, but in some places nearly as numerous as the fox. Surprisingly little is known about wildcat populations or the animals' habits. Gamekeepers were responsible for cutting their numbers drastically in the eighteenth and nineteenth centuries, but with the decline of the big shooting estates after World War II the cat seems to have increased in number and range. Its main food seems to be rabbits and voles but like a domestic cat, it will eat many items which come its way. It weight varies from 7 to 11lb (3–5kg). It has one litter of two to four kittens in May and usually another in August; gestation takes sixty-three days (as compared with fifty-eight in the domestic cat). The wildcat is fully grown at ten months.

Pine martens and polecats are found in small

A red deer stag stands in early winter snow in the Scottish Highlands. Britain's largest native deer and largest land mammal, red deer are widespread in the Highlands, though colonies also exist in the Lake District, south-west England, the Breckland of East Anglia and the New Forest. The stag's fine pair of antlers, grown between April and July, is shed the following February.

243

numbers in the hills. The pine marten is very seldom seen, and then only by chance. It lives in the wild, rugged central and western Scottish Highlands, and is more a forest animal, preferring coniferous forest, although dens have been found on open, rocky hillsides. Martens are acrobatic, arboreal hunters and are athletic enough to catch squirrels in the treetops; however, studies have shown that a high proportion of their diet consists of field voles. They occasionally come to bird tables, and may raid chicken runs. So, too, do polecats, their name being perhaps derived from the French *poule*, a chicken. Polecats were not originally upland mammals, but because they were extensively persecuted their range is now limited to the mountain valleys of central Wales. The population may be expanding, but even so, they are extremely elusive and the chance of ever getting even a casual glimpse of one is slight.

The pine marten's weight varies from 2 to 3½lb (1 to 1.5kg) and it has one litter of two to three kittens in late March or April. Polecats weigh about 3lb (1.25kg) and produce two litters per year.

The two smaller mustelids, the stoat and the weasel, are much more widespread and are often seen in hill country, usually hunting along the base of a stone wall, their dens in holes deep beneath the stones. They are naturally inquisitive and may be persuaded to come quite close, especially to an imitation of a rabbit squeal.

Small herds of 'feral' wild goats are found in many hill districts, especially in Scotland, but also in Snowdonia and the Cheviots. They can be almost any colour or combination of colours from black, white, grey and varying shades of brown to piebald and skewbald.

CONCLUSION

THE catalogue of environmental damage which has appeared in this book makes dismal reading. Overpopulation is responsible for our excessively competitive consumer society which is precipitating potentially lethal changes in the atmosphere, and overpopulation drives the urban developer to use every last inch of available land for a few more houses, destroying irreplaceable natural habitats as he goes.

Potentially the most serious threats are posed by acid rain, the so-called 'greenhouse effect', and the destruction of the ozone layer, all three of which are caused by atmospheric pollution. Burning carbon-based fuels – mainly fossil fuels like coal, oil and petrol – produces carbon dioxide; it also produces sulphur dioxide, and various nitrogen oxides; widespread damage is caused when these latter gases condense from the atmosphere and fall as 'acid rain'. The concentration of carbon dioxide in the upper atmosphere has increased by 25 per cent in the last twenty years. Trees absorb carbon dioxide and could well ease this particular problem if enough of them are left standing – yet we continue to cut them down at a huge rate. The concentration of methane has doubled. The presence of these two gases in the upper atmosphere has the effect of trapping solar heat and as a consequence the atmosphere we live in may become considerably warmer. There are various predictions as to what the consequences might be should the earth's climate warm up significantly, but some foresee a melting of the ice-caps, then widespread flooding and wholesale crop failure.

The ozone layer round the earth is of crucial importance because it reduces the effect of ultraviolet light from the sun. Damage to the ozone layer is caused by chemicals used in most fridges and freezers, and in some aerosols, chemicals known as CFCs – chloro-fluoro-carbons. There is a huge area over Antarctica which is already damaged. If the ozone layer is breached, more harmful ultraviolet light reaches earth and this can cause crop damage and skin cancers. All the scientific evidence shows that we are nearing the point of no return; small wonder that in 1988 the British government started to make 'green' noises – action, of course, is quite another matter.

In Britain the more specific and most intractable problems are acid rain, chemical and oil pollution of the seas, nuclear waste and its disposal, intensive farming – including the excessive use of toxic chemicals – and suburban development, while inner cities decay; also traffic pollution, and household waste and sewage disposal; and last, but by no means least, widespread conifer plantation and the pressure of visitor numbers on the best and therefore the most precious wild places.

Acid rain is not just 'dirty rain on the landscape'. Every year millions of tonnes of acid pollutants, from the burning of fossil fuels, enter the atmosphere. Sulphur dioxide is the most significant producer of acid rain; 20 million tonnes of SO_2 enter the atmosphere of Western Europe every year. Nitrogen oxides have risen sharply since 1960, because of the rise of the motor vehicle, so that we have on average 9 million tonnes of them in our atmosphere. Power stations are responsible for a staggering 71 per cent of SO_2 and 37 per cent of NO emissions.

Acid rain damage abroad is greater than in Britain because of the prevailing south-west winds. Of Sweden's 90,000 lakes 18,000 are acidified to a 'dead' level, with no fish stocks. In Norway half the stocks of freshwater fish, including salmon and trout, are regarded as lost. Over 1¼ million acres (500,000ha) of German forests have died, and the German government anticipates the death of all of the Black Forest by the year 2000 – just eleven years away. Even Scottish forests are affected, with some areas showing acid damage. However, Germany has already spent £6 billion cleaning up power stations and 80 per cent of its power stations are 'clean'. Britain on the other hand, proposes to spend a mere £0.9 billion by 1997, and 'clean' just three of its forty-plus fossil-fuel power stations.

Sea pollution hit the headlines in 1988 when seals started dying in huge numbers in the North Sea. Burning toxic wastes at sea is a primitive method of waste disposal and extremely harmful, as considerable quantities of toxic residues are left, especially in the surface layers of the sea. The North Sea is the only place in the western world where this is allowed; and it is further abused by huge concentrations of toxic chemicals which are also poured into it daily. Britain alone disposes of over 14 million tons of contaminated waste, including over 4,800 tons of heavy metals, plus over 10 million tons of sewage every year, all of it into the North Sea – and the Rhine deposits much more than this. Recent large-scale eco-disasters in the North Sea include the seal deaths, huge blooms of algae and depleted fish stocks; sea-birds have failed to raise chicks, whole populations dying because the supplies of small fish such as sand eels have been destroyed; and dolphins and porpoises have disappeared. All of these creatures are also threatened by the fishing industry, when they are drowned by the hundred in new nylon nets. There are still strident calls from fishermen for seals to be heavily culled, as they believe the seals compete too avidly for the fish available. In fact it is the fishing industry itself which has been responsible for the wholesale over-fishing of several species, almost wiping out some of the fish stocks altogether.

Pollution from oil subjects sea life to yet further attack. Strong controls are exercised round Britain's offshore oilfields and harbours but accidents are inevitable. Much more reprehensible is the illegal and deliberate cleaning out of ships' tanks by rogue oil-tankers, a practice carried out simply to save money.

Nuclear-waste disposal presents a real and irreversible problem; where does one put x tons of radioactive material which may still be lethal many hundreds of years in the future? 'Not in my backyard' is a perfectly understandable answer, but does nothing to solve the problem.

Modern agricultural practices are equally to blame. In the last thirty years many of Britain's farmers have been responsible for turning huge areas of farmland into highly productive but ecologically sterile prairies. Or rather, over-productive, because now we have 'mountains': butter mountains, cereal mountains and beef mountains, all of it produce which is surplus to requirement – in 1985 £2 million per day was spent by the EEC on storing these items. Over-production of fruit and vegetables means that tons of produce are destroyed every day; this is estimated to be at such rates as 37lb (17kg) of cauliflowers, 119lb (54kg) of tomatoes and 2,448lb (1,110kg) of apples thrown out every minute of each working day.

We are the ordinary people, the silent majority: is there anything we can do, bearing in mind that everybody wants heating, cars, fridges, paper and all the other luxuries we have become accustomed

PREVIOUS PAGE
An industrial landscape can look almost beautiful in the sunset but this belies the harm industry is doing, daily adding to the levels of pollution in our environment. Man has the technological knowledge and expertise to save this planet, just as he employs the technology which is destroying it.

Beech woods in autumn – a
stunning image of our
wildlife heritage.

to? Yes, there is: bombard your member of parliament and ministers with letters about environmental issues – if they don't hear from you they think you don't care. Join your country wildlife trust and take an active part. Join the Royal Society for the Protection of Birds which helps save threatened habitats on a national scale. Be aware of the urban developer's local developments, and protest – with reasoned argument – to your local planning authority; do not tolerate 'creeping suburbia'. Above all, cut your own pollution levels: stop using aerosols, avoid garden pesticides, recycle glass and paper and tins if you can, don't waste energy and don't drop litter – especially plastics, which may still be where you left them 100 years from now. And invest your money in a catalysing car exhaust which removes harmful gases.

Today more than ever before man is destroying the beautiful natural world in which he lives – a world filled with a multiplicity of colour, symmetry, grace and power, far in advance of anything the human world has ever been able to manufacture. Man has the technological knowledge and expertise to save this planet, just as it is his technology which is destroying it – but does he have the will? The world and its wildlife is at a crossroads and the next few years will reveal whether or not our grandchildren will still have a beautiful environment in which to live as we pass into the twenty-first century. It rests with every one of us to ensure a healthy future for the earth, our wildlife and our own environment.

BIBLIOGRAPHY

A book such as this obviously draws upon many sources of information including many specialist reference books. The author would like to acknowledge the use of the following books which were consulted during the preparation of this book. This list also includes many good field identification guides.

ALLABY, M. *The Woodland Trust Book of British Woodlands* David & Charles 1986

ANGEL, H. *The Natural History of Britain & Ireland* Michael Joseph 1981

ANGEL, H. *Heather Angel's Countryside* Michael Joseph/Rainbird 1983

ANGEL, H. and WOLSELEY, P. *Waterways and Waterlife of Great Britain* Peerage Books 1986

ARLOT, N., FITTER, R. and FITTER, A. *The Complete Guide to British Wildlife* Collins 1981

ATTENBOROUGH, D. *Life on Earth* Collins/BBC 1979

ATTENBOROUGH, D. *The Living Planet* Collins/BBC 1984

BAKER, R. (ed) *The Mystery of Migration* Macdonald

BARRET, J. and YONGE, C. M. *Collins Pocket Guide to the Seashore* Collins 1972

BLAMEY, M. and GREY-WILSON, C. *The Illustrated Flora of Britain and Northern Europe* Hodder & Stoughton 1989

BROOKES, B. *BNA Guide to Mountain and Moorland* Crowood Press 1985

BROOKS, M. and KNIGHT, C. *A Complete Guide to British Butterflies* Jonathan Cape 1982

BURTON, R. *Carnivores of Europe* Batsford 1979

CADMAN, A. *Dawn, Dusk and Deer* Country Life 1966

CHINERY, M. *A Field Guide to the Insects of Britain and Western Europe* Collins 1974

CHINERY, M., and TEAGLE, W. G. *Wildlife in Towns and Cities* Country Life Books 1984

CLARK, M. *Badgers* Whittet Books 1988

COLEBOURN, P., and GIBBONS, B. *Britain's Natural Heritage* Blandford Press 1987

COOPER, A. *Secret Nature of Britain* BBC 1989

CORBET, G. B., and SOUTHERN, H. N. (eds) *The Handbook of British Mammals* Blackwell 1977

CORBET, P. S., LONGFIELD, C., and MOORE, N. W. *Dragonflies* Collins New Naturalist no 41 1960

CRAMP, S., BOURNE, W. R. P., and SAUNDERS, D. *The Seabirds of Britain and Ireland* Collins 1974

CRAWFORD, P. *The Living Isles* BBC 1985

DARLING, F., and BOYD, W. *The Highlands and Islands* Collins

DICKINSON, C., and LUCAS, J. *Encyclopaedia of Mushrooms* Orbis 1979

DURMAN, R. (ed) *Bird Observatories in Britain and Ireland* T. and A. D. Pyser

FITTER, R., FITTER, A. and BLAMEY, M. *The Wild Flowers of Britain and Northern Europe* Collins 1974

FITTER, R., and MANUEL, R. *Field Guide to Freshwater Life* Collins 1986

GIBBONS, B. *Wildlife of the British Isles* Country Life 1987

GILMOUR, J. and WALTERS, M. *Wild Flowers* Collins New Naturalist 4th ed 1969

GOODDEN, R. *British Butterflies: A Field Guide* David & Charles 1978

GOODERS, J. *Field Guide to the Birds of Britain and Ireland* Kingfisher Books 1986

GROUNDS, R. *Ferns* Pelham Books 1974

HAMMOND, N. (ed.) *RSPB Nature Reserves* RSPB 1983

HARRIS, S. *Urban Foxes* Whittet Books 1986

HARRIS, S. *The Secret Life of the Harvest Mouse* Hamlyn 1979

HARRISON-MATTHEWS, L. *Mammals in the British Isles* Collins New Naturalist no 68 1982

HEINZEL, H., FITTER, R. S., and PARSLOW, J. *The Birds of Britain and Europe* Collins 1972

HEWER, H. R. *British Seals* Collins New Naturalist no 57 1974

HIGGINS, L. G., and RILEY, N. D. *A Field Guide to the Butterflies of Britain and Europe* Collins 1980

HOLM, J. *Squirrels* Whittet Books 1987

JAHNS, H. M. *Collins Guide to the Ferns, Mosses and Lichens of Britain and Northern and Central Europe* Collins 1983

LANGE, M., and BAYARD-HORA, F. *Collins Guide to Mushrooms and Toadstools* Collins

LAWRENCE, M. J., and BROWN, R. W. *Mammals of Britain – their Tracks, Trails and Signs* Blandford 1973

LEE, B. *BNA Guide to Fields, Farms and Hedgerows* Crowood Press 1985

LLOYD, J. G. *The Red Fox* Batsford 1980

LOCKLEY, R. M. *Grey Seal, Common Seal* André Deutsch

LOFGREN, L. *Ocean Birds* Croon Helm 1984

LOUSLEY, J. *Wild Flowers of Chalk and Limestone* Collins New Naturalist no 16 1950

LOVEGROVE, R., and SNOW, P. *River Birds* Columbus Books 1984

McCLINTOCK, D., and FITTER, R. S. R. *Collins Pocket Guide to Wild Flowers* Collins

MEAD, C. *Bird Migration* Country Life Books 1983

MILES, H. *The Track of the Wild Otter* Hamish Hamilton 1984

MILES, P. M. and H. B. *Woodland Ecology* Hulton Educational 1968

MITCHELL, A. *A Field Guide to the Trees of Britain and Northern Europe* Collins 1974

MITCHELL, A. *Broadleaves* Forestry Commission 1985

MITCHELL, A. *Conifers* Forestry Commission 1985

MORRIS, P. *Hedgehogs* Whittet Books 1983

MORRIS, P. (ed) *The Natural History of the British Isles* Country Life Books 1979

MUIR, R. and N. *Hedgerows* Michael Joseph 1987

NEAL, E. *The Natural History of Badgers* Croon Helm 1986

NELSON, B. *Seabirds – their Biology and Ecology* Hamlyn 1979

NÉTHERSOLE-THOMPSON, D. *Highland Birds* HIDB 1971

NICHOLS, D., and COOKE, J. *The Oxford Book of Invertebrates* OUP 1971

ODDIE, B. *Bird Watching* Macmillan 1988

OGILVIE, M. A. *Wild Geese* T. and A. D. Poyser 1978

O'TOOLE, C. *The Encyclopaedia of Insects* Collins 1977

POLLARD, E., HOOPER, M., and MOORE, N. *Hedges* Collins New Naturalist no 58 1974

PETERSEN, R., MOUNTFORT, G., and HOLLOM, P. *A Field Guide to the Birds of Britain and Europe* Collins 1983

PHILLIPS, R. *Wild Flowers in Britain* Pan Books 1977

PHILLIPS, R. *Trees in Britain* Ward Lock 1978

PHILLIPS, R. *Mushrooms* Pan Books 1981

PHILLIPS, R. *Grasses, Ferns, Mosses and Lichens of Great Britain and Ireland* Pan Books

PRATER, A. J. *Estuary Birds of Britain and Ireland* T. and A. D. Poyser 1981

PRESTT, I. *British Birds: Lifestyles and Habits* Batsford 1982

PRIOR, R. *Deer Watch* David & Charles 1987

RACKHAM, O. *The History of the Countryside* J. M. Dent 1986

RATCLIFFE, D. (ed) *A Nature Conservation Review* NCC 1977

RATCLIFFE, D. *Highland Flora* HIDB 1977

READER'S DIGEST ASSOCIATION *The Everchanging Woodlands* 1984

READER'S DIGEST ASSOCIATION *The Evergreen World* 1986

READER'S DIGEST ASSOCIATION *Seas and Islands* 1986

RICHARDS, A. J. *British Birds: A Field Guide* David & Charles 1979

RICHARDSON, P. *Bats* Whittet Books 1985

ROBINSON, R. K. *The Ecology of Fungi* EUP 1967

SOOTHILL, E., and FAIRHURST, A. *The New Field Guide to Fungi* Michael Joseph

SOPER, T. *The Bird Table Book* David & Charles 1989

SOPER, T. *The National Trust Guide to the Coast* Webb Y. Boyer 1984

SOPER, T. *Discovering Animals* BBC 1985

SOPER, T. *Go Birding* BBC 1988

SPARKS, J. *The Discovery of Animal Behaviour* Collins/BBC 1982

STREETER, D., and RICHARDSON, R. *Discovering Hedgerows* BBC 1982

SUMMERHAYES, V. S. *Wild Orchids of Britain* Collins

TAYLOR-PAGE, F. *Field Guide to British Deer* Blackwell 1971

THOMPSON, G., COLDREY, J., and BERNARD, G. *The Pond* Collins 1984

TWEEDIE, M. *Insect Life* Collins 1977

VAN DEN BRINK, F. H. *A Field Guide to the Mammals of Britain and Europe* Collins

WHITEHEAD, G. K. *The Deer of Great Britain and Ireland* Routledge and Kegan Paul

WHITLOCK, R. *The Oak* G. Allen & Unwin 1985

WILLIAMS, C. B. *Insect Migration* Collins New Naturalist 1958

WOOD, R. M. *On the Rocks: A Geology of Britain* BBC Books

YALDEN, D. W., and MORRIS, P. A. *The Lives of Bats* David & Charles 1975

See also leaflets published by the Countryside Commission, Forestry Commission, Nature Conservancy Council, National Trust, National Trust for Scotland, national parks and country wildlife trusts.

ACKNOWLEDGEMENTS

A book such as this seems to require half a lifetime of information gathering and practical experience.

I am indebted to all those people who have shared their experience with me during the years of exploration, observation and recording. Without them I should never have learned about the variety of wonderful wildlife all about us.

I am particularly grateful to Dr Derek Valden of the Zoology Department, Manchester University, an old friend of many years, who kindly vetted and edited the text.

Thanks to Tracey May and Jane Rowe, my editors, and to the rest of the team at David & Charles, while Frances Kelly steered the book in the right direction. Thanks too, to Hazel Cooper, who produced a faultless typescript from my usually illegible handwriting. And lastly to my wife Joan who kept me right up to schedule with constant practical assistance.

USEFUL ADDRESSES

Amateur Entomological Society
355 Hounslow Rd, Hanworth, Middx TW13 5JH

Botanical Society of British Isles
68 Outwood Rd, Loughborough, Leicester

British Butterfly Conservation Society
Tudor House, Quorn, Leicester LE12 8AD

British Deer Society
Church Farm, Lower Basildon, Reading, Berks RG8 9NH

British Entomological Society
74 South Audley St, London W14 5FF

British Trust for Conservation Volunteers
10–14 Duke Street, Reading, Berks RG1 4RU

British Trust for Nature Conservation
36 St Mary's St, Wallingford, Oxford OX10 0EU

British Trust for Ornithology
Beech Grove, Station Rd, Tring, Herts HP23 5NR

Countryside Commission
John Dower House, Crescent Place, Cheltenham, Glos GL50 3RA

Field Studies Council
9 Devereux Court, The Strand, London WC2R 3JR
(runs nine residential centres and one day centre)

Flora & Fauna Preservation Society
c/o Zoological Society, Regent's Park, London NW1 4RY

Forestry Commission
231 Corstorphine Road, Edinburgh EH12 7AT

Mammal Society
Dexter House, 2 Royal Mint Court, London EC3N 4XX

National Trust
42 Queen Anne's Gate, London SW1H 9AS
(plus fifteen regional offices)

National Trust for Scotland
'Suntrap', 43 Gogarbank, Edinburgh EH12 9BY

Nature Conservancy Council
Northminster House, Peterborough PE1 1UA
(plus fifteen regional offices)

Royal Entomological Society of London
41 Queens St, London SW7 5HU

Royal Society for Nature Conservation
The Green, Nettleham, Lincolnshire LN2 2NR
(co-ordinates county wildlife trusts)

Royal Society for the Protection of Birds
The Lodge, Sandy, Beds SG19 2DL
(plus ten regional offices)

Scottish Wildlife Trust
25 Johnston Terrace, Edinburgh EH1 2NH

Wildfowl Trust
Slimbridge, Gloucester GL2 7BT

Woodland Trust
Westgate, Grantham, Lincoln NG31 6LL

World Wide Fund for Nature
Panda House, 11–13 Ockford Rd, Godalming, Surrey GU7 1QU

INDEX